PEDESTRIAN SAFETY

PT-112
Pedestrian Safety

Other SAE books of interest:

Automotive Safety Handbook
By Ulrich W. Seiffert and Lothar Wech
(Order No. R-325)

Pedestrian Safety

PT-112

Edited by
Daniel J. Holt

Published by
Society of Automotive Engineers, Inc.
400 Commonwealth Drive
Warrendale, PA 15096-0001
U.S.A.
Phone (724) 776-4841
Fax: (724) 776-5760
www.sae.org
January 2004

For permission and licensing requests contact:

SAE Permissions
400 Commonwealth Drive
Warrendale, PA 15096-0001-USA
Email: permissions@sae.org
Fax: 724-772-4891
Tel: 724-772-4028

Global Mobility Database®

All SAE papers, standards, and selected books are abstracted and indexed in the Global Mobility Database.

For multiple print copies contact:

SAE Customer Service
Tel: 877-606-7323 (inside USA and Canada)
Tel: 724-776-4970 (outside USA)
Fax: 724-776-1615
Email: CustomerService@sae.org

ISBN 0-7680-1342-9

Library of Congress Catalog Card Number: 2003115015
SAE/PT-112
Copyright © 2004 SAE International

Positions and opinions advanced in this publication are those of the author(s) and not necessarily those of SAE. The author is solely responsible for the content of the book.

SAE Order No. PT-112

Printed in USA

Preface

Pedestrian Safety
By Daniel J. Holt

A recent research report released by the U.S. Department of Transportation's National Highway Traffic Safety Administration (NHTSA) has stated that almost 175,000 pedestrians died on U.S roadways between 1975 and 2001. It was also noted in the report that 12% of all deaths related to motor vehicle crashes in the country are pedestrian fatalities.

In 2001 there were 42,116 traffic deaths in the U.S. with 12% or 4882 of these fatalities being pedestrians. Earlier reports have shown that in 1999 pedestrians accounted for 85,000 injuries and 5000 fatalities in traffic accidents in the U.S. Similar reports on Western Europe for the same period revealed that 7000 pedestrians were killed and 74,000 were injured. In Japan there were 85,000 pedestrians were injured and 3000 pedestrian were killed in 1999 due to traffic accidents.

What can vehicle manufacturers do to reduce pedestrian fatalities?

Most of the safety technology to date in vehicles has been applied to protect the occupants in the vehicle. Automobile manufacturers are now examining ways the automobile can be designed to help reduce fatalities and injuries when a pedestrian and vehicle meet during an impact.

Research is being directed on two major fronts. Engineers are looking at methods to sense the presence of pedestrians and warn drivers of their location relative to the vehicle. The NHTSA statistics show that 18% of the pedestrian fatalities occur between midnight and 6 a.m. while 46% occur between 6 p.m. and midnight.

The April 2003 NHTSA report had some other interesting data on pedestrian fatalities between 1998 and 2001. A look at where the fatalities occurred revealed that 78% occurred at non-intersections, 64% occurred on urban highways, 44% occurred on roadways without crosswalks, and 18% were hit-and-run crashes.

If one takes a look at the statistics, it appears that almost 66% of the pedestrian fatalities occur between 6 p.m. and 6 a.m. or during the night or late evening. This has led the vehicle manufacturers to examine night vision systems to allow the driver to see a pedestrian crossing the road under less than ideal lighting conditions. With 64% of the pedestrians killed on urban highways, night vision and pedestrian detection systems could help reduce fatalities by allowing the driver to avoid the pedestrian.

Even with the best pedestrian sensor systems, night vision systems, and improved lighting systems, some impacts between a pedestrian and a vehicle may be unavoidable. The second front that the automobile manufacturers are examining is designing the vehicle to reduce the injury level of a pedestrian during a crash.

The automobile companies examined the type of impact a vehicle would have with a pedestrian by examining three basic areas using test forms to study impacts to the head, upper leg, and lower leg.
The specialized crash test forms can duplicate impacts to various body parts and help researchers to analyze the impacts and thereby design safety systems.

The NHTSA report also spelled out that more than a fifth of all children between the of ages 5 and 9 killed in traffic accidents were pedestrians. However, the age group with the highest rate of pedestrian fatalities were those 70 and above. They accounted for 20% of the fatalities. Thus, the vehicle manufacturers must design vehicles that can help not only adults of various age groups to survive an impact between them and a vehicle but also children that are smaller than most adults.

Air bags inside the vehicle have proved to be useful in saving lives since they were introduced in the 1980s. Some vehicle manufacturers have examined using external air bags to help reduce the impact between a pedestrian and a vehicle.

European regulations

The European Union (EU) has set forth that all member countries must meet new HIC (head injury criteria) starting in July 2005. The agreement stipulates that vehicle manufacturers monitor the HIC values of all new models. The EU has established a series of tests that will be conducted to ensure that the violence to a pedestrian's head does not exceed certain levels when the head, with a speed of 35 km/h (22 mph) hits the vehicle. This test criterion corresponds to a vehicle moving at 40 km/h (25 mph). The EU test levels are 1000 HIC for two thirds of the hood surface and 2000 HIC for the remaining one third of the hood. The HIC is an internationally accepted acceleration-based measurement for violence against the head. HIC values under 1000 imply that the risk of a life-threatening injury is 15% or less while at 2000 the fatality risk is almost 90%.

Vehicle manufacturers worldwide are concerned with the levels of pedestrian fatalities with vehicle and are examining methods to reduce the fatalities.

TABLE OF CONTENTS

The Virtual Stiffness Profile –
A Design Methodology for Pedestrian Safety

Alexander Droste, Pádraig Naughton and Peter Cate
Dow Automotive[1]

ABSTRACT

European car manufacturers and suppliers are currently stepping up the effort to develop solutions to meet pedestrian safety requirements, which will come into effect, starting in 2005. Numerous concepts, both active and passive, are being investigated to fulfil the pedestrian safety specifications, in addition to the many other limitations imposed on the front end of the car. All of them deal with the topic of energy absorption. Here, an approach to achieving a passive solution will be presented, describing the development of the 'Virtual Stiffness Profile' (VSP) to help identify the optimum balance of engineering and styling to meet the requirements. In this paper, specific emphasis is placed on the lower leg impact.

INTRODUCTION

The needs of pedestrian safety are different from current requirements such as low speed or insurance impacts. To fulfil both traditional and pedestrian safety requirements, design changes are needed to find a good balance. However, design limitations are imposed in order to conserve the styling and aesthetics of the front-end, which define the image and often handling/aerodynamics of the car. Thus, numerous boundary conditions, both mechanical and non-mechanical, should be taken into account during the implementation of pedestrian safety solutions

PEDESTRIAN SAFETY – GENERAL OVERVIEW

Pedestrian safety is moving into a new phase, where the requirements are being considered seriously in the development of new vehicles. Testing according to EuroNCAP is being conducted and published according to EEVC proposals and a modification of these tests has been confirmed according to ACEA proposals for voluntary implementation over a period of time and in a stepwise fashion. Rather than full pedestrian dummy testing, the common approach in Europe will remain as component tests representing the main areas of injury in a pedestrian impact with a car front, namely lower- and upper leg, child- and adult head.

LEGISLATION STATUS

The European Commission accepted the voluntary commitment of the European Automobile Manufacturers Association (ACEA) on pedestrian protection on 26th of Nov. 2001. In addition to several short term actions, it has been agreed that from the 1st of July 2005 all new vehicle types shall comply with a first set of passive safety measures, recommended by the Commission's Joint Research Center [1, 2]. From 2010 all new vehicle types shall meet the limits defined by the European Enhanced Vehicle Safety Committee (EEVC) or equivalent measures. A sketch summarising the ACEA/EEVC requirements is shown in figure 1.

Given the challenges associated with meeting even the initial limits, this paper provides a methodology whereby the often-conflicting requirements, which influence the structure of the car front, can be taken into account and a successful compromise reached.

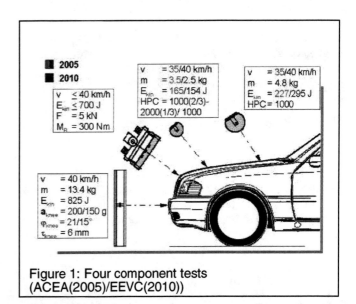

Figure 1: Four component tests (ACEA(2005)/EEVC(2010))

---

[1] Dow Automotive is a business unit of The Dow Chemical Company and its subsidiaries

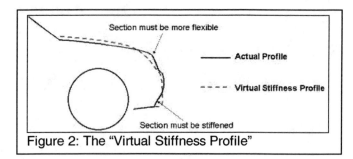

Figure 2: The "Virtual Stiffness Profile"

THE VIRTUAL STIFFNESS PROFILE

By introducing a regulation or, in this case, a voluntary commitment, boundary conditions are indirectly imposed on the car. Since every car type is different in the use of materials and geometric shapes, the definition of the dummy components and the limits on their behaviour during impact are the fixed parts in this kind of boundary value problem.

From the combination of all four requirements, in addition to all the current limitations, the concept of an ideal profile of a car front compared to a real car front design has been introduced [3, 4]. This "virtual stiffness profile" (VSP) defines the stiffness distribution, which is optimised to absorb the various impact energies while maintaining the original styling wherever possible (figure 2). In all four cases the contact and rebound behaviour of the dummy components provide information on the best behaving car front geometry.

BEST CASE PROFILES

In a first step, assuming ideal energy absorption, the best case of the shape for each counterpart of the contacting component is created. The combination of these best-case profiles results in the ideal VSP, against which the real behaviour will be optimised through stiffness distribution and, if possible, with limited geometric shape changes.

The real crash situation of a moving car impacting a moving or standing object or person is represented by a moving component hitting a stationary car, using the appropriate amount of energy. In all cases this pre-contact energy is a pure kinetic energy E_{kin} which can be described by

$$E_{kin} = 0.5\, m\, v^2 \qquad (1)$$

where m is the mass of the component and v is the magnitude of the velocity vector. To absorb this energy during impact with the car, the work done W is described by

$$W = \int_{\Delta x} F \cdot dx \qquad (2)$$

where x is the intrusion into the car and F is the resisting force. This force can be derived from the Newtonian equation of motion

$$\vec{F} = m\vec{a}\,, \qquad (3)$$

with the acceleration vector $\vec{a}$. To fulfil the conservation of energy, the kinetic energy is related to the work done W by

$$\Delta E_{kin} + \Delta W = 0\,. \qquad (4)$$

Equations (1)–(4) can be used to estimate the minimum required deformation $\vec{u} = \Delta\vec{x}$ of the car components, given a limitation on the force or acceleration which is allowed on the impactor.

Ideal energy absorption

Depending on the contact conditions the kinetic energy is partially or fully transferred into internal energy through deformation or heat through friction. In the following it is assumed that the energy dissipated through friction is negligible. Thus, only the deformation of the components is taken into account. Because the deformation process can be divided into reversible (index: rev) and irreversible (index: irr) parts this condition holds also for the system energy E:

$$E = E_{rev} + E_{irr}\,. \qquad (5)$$

This equation includes the material behaviour of the impacting component as well as the car components influencing the absorption performance. The equation of the deformation work absorbed by the material W_{EA}, also called strain energy, described by

$$W_{EA} = \int_{\Delta\varepsilon} \sigma \cdot d\varepsilon\,, \qquad (6)$$

with the stress σ and the deformation ε, indicates the optimal material parameters representing the general behaviour of the car front.

For simplification the creation of ideal energy absorption is considered from a one-dimensional point of view. With this in mind, figure 3 shows the stress-strain behaviour of an ideal absorbing material, assuming a limitation on the stress, force or acceleration level allowed, a uniform contact area and goal of minimising the deformation. The energy absorption efficiency η is then defined by

$$\eta = \frac{\int_{0}^{\varepsilon} \sigma \cdot d\varepsilon}{\sigma_{max}\, \varepsilon_{max}}\,. \qquad (7)$$

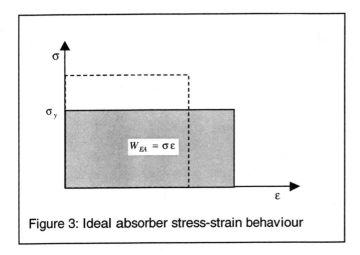

Figure 3: Ideal absorber stress-strain behaviour

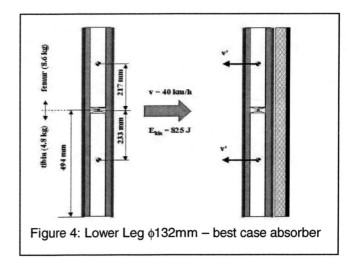

Figure 4: Lower Leg ϕ132mm – best case absorber

Here, a perfectly-plastic behaving material, also called ideal plastic, indicated by the yield stress σ_y, shows the most effective ($\eta = 1$) absorption behaviour, as the area below the stress-strain curve is maximised.

Depending on the boundary conditions, an optimum of the allowed stress and, more importantly for the design space, the strain is related to the maximum allowed acceleration.

In the following discussions the summarising equation for the one-dimensional problem

$$E_{kin} = \frac{m}{2}v^2 = Fu = mau = \sigma_y A_0 \varepsilon l_0 = W_{EA} \qquad (8)$$

will be used to find ideal values of the intrusion of the components into the car. In addition to the variables described above, the contact area A_0 and the initial thickness l_0 of the ideal absorber are included.

Lower Leg

The Legform requirements impose restrictions on the maximum deceleration, the maximum knee-bending angle and the maximum knee-shear deflection. The bending angle and the shear are strongly related to the relative deforming behaviour of the different parts of the front-end, which dictate the kinematics of the legform. Assuming that all of the kinetic energy of the legform is absorbed and considering the legform geometry and impact conditions, a vertical car front geometry is the ideal shape. This yields 0° bending angle and 0 mm shear in impact and rebound, when velocities of the tibia and the femur centres of gravity (COG's) are maintained at equal levels. For the deceleration of the knee, the vertical profile offers the biggest contact surface and is, for the required conditions, the best case (see figure 4).

Considering the boundary conditions for the lower leg component of 2010 (see figure 1), the mass of the leg, the initial kinetic energy of the leg and the maximum allowed acceleration (a=150 [g] = 1471.5 m/s²) result in a minimum design space needed for the absorption,

$$u = \frac{E_{kin}}{F} = \frac{E_{kin}}{ma} = \frac{825\,J}{13.4\,kg\,150\,9.81\,m/s^2} = 42\,mm . \qquad (9)$$

In relation to the leg geometry, the contact area can be defined by a simplified approach assuming 112 mm as the width of the contact area. The height is variable to a maximum of the complete leg height ($h_l = 876\,mm$). Depending on this height, the required yield stress value of the ideal absorber can be derived by

$$\sigma_y = \frac{19718.1\,N}{112\,h_l\,mm^2} . \qquad (10)$$

For the full leg height a value for the yield plateau of $\sigma_y = 0.2\,MPa$ is needed.

These values are only a first approximation, not taking into account the non-linear influences of the deformation of the leg-foam as well as the change of the contact area A throughout the impact time. It is assumed that $A = A_0$ holds. Further development of the virtual stiffness profile (VSP) and the relationship to the real profile (RSP) of the front of the car is shown below. Here, the influence of dynamic impact, changing contact area, component foam behaviour and the kinematics of the legform, where the kinetic energy is not completely absorbed, are considered.

An introduction to the approach for the upper leg and head form is presented in the following sections, but due to the amount of information, detailed descriptions will be included in future papers.

Upper Leg

In the case of the upper leg component, the 'upper legform to bumper' test and the 'upper legform to bonnet edge' tests may be performed. The 'upper legform to bumper' test is suggested if the lower bumper reference

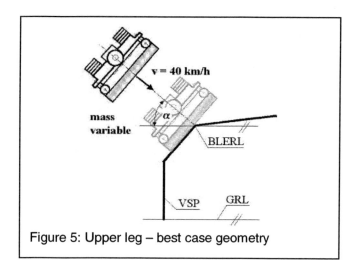

Figure 5: Upper leg – best case geometry

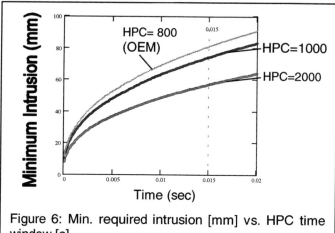

Figure 6: Min. required intrusion [mm] vs. HPC time window [s]

line is greater than 500 mm. Also in this case, the vertical profile minimises the bending moment and a full contact surface best distributes the contact force. The bonnet leading edge reference line is limited by the 1000 mm wrap around distance.

For the upper leg the influence of the component, as well as the position, has to be taken into account. Thus, most of the energy absorption is related to the deformation of the upper leg-foam itself. For an ideal profile it is assumed initially that at least the lower half of the upper-leg-component is in full contact with the car, and the COG axis hits the part ideally as shown in figure 5.

In the case of the upper leg component, the 'legform to bonnet edge' test is only required if, based on certain geometry conditions, the kinetic energy of the upper legform is greater than 200 Joules. No test is required if the kinetic energy is less than 200 Joules. The shape window for this case, is between a bonnet leading edge height (BLEH) of 650 mm with a bumper lead (BL) of 50 mm and a bonnet leading edge height of 725 mm with a bumper lead of 350 mm. This means that with some freedom in design, it is possible to eliminate the need for an upper legform test. Further detail of the upper leg requirements and conditions can be found in the ACEA voluntary commitment [2].

Child and adult head

For the head components, calculation of the deflection needed to absorb the energy becomes as complex as for the lower leg. The head performance criterion HPC (also known as the head injury criterion HIC) as defined by the equation

$$HPC = \max\left\{\frac{1}{(t_2 - t_1)}\int_{t_1}^{t_2} a(t)dt\right\}^{2.5}(t_2 - t_1) \qquad (11)$$

is an optimisation of the contact time window $(t_2 - t_1)$ where the area under the acceleration curve related to

the time window itself, to the power of 2.5, is maximised over a time window between 0 and 15 ms.

Assuming an energy absorption mechanism, which yields an ideal constant acceleration over time, equation (11) can be simplified as

$$HPC = a^{2.5}(t_2 - t_1) , \qquad (12)$$

where the time window $(t_2 - t_1)$ is less than or equal to 15 ms. The intrusion required to absorb all the kinetic energy is then shown in figure 6, as a function of time interval and HPC value allowed.

The intrusion required is reduced, when the acceleration levels are increased to high levels for a short period of time. However, the acceleration levels for short time intervals become quite high (figure 7) and the stiffness required to achieve this behaviour becomes unrealistic.

During the head impact event some of the initial translational kinetic energy is transformed into rotational kinetic energy, which manifests itself in the rolling of the head on the deforming bonnet, and need not be absorbed. The contact area changes throughout the impact event and the foam of the head form also absorbs some of the energy. All of these factors make

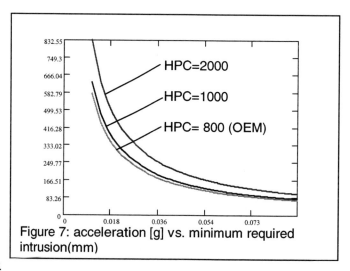

Figure 7: acceleration [g] vs. minimum required intrusion(mm)

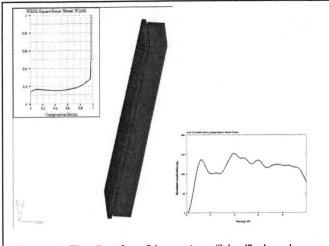

Figure 8: Flat "leg form" impact on "ideal" absorber. Square stress-strain behaviour results in a relatively flat acceleration curve, allowing for dynamic effects.

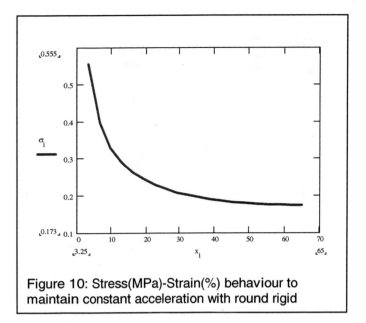

Figure 10: Stress(MPa)-Strain(%) behaviour to maintain constant acceleration with round rigid

complete analytical prediction of the behaviour extremely difficult.

VSP GENERATION

From the best case profiles it is possible to construct an ideal geometric profile optimising all four best case scenarios into one, final virtual stiffness profile. The development of the approach for the lower legform is shown in some detail below.

Constant contact area – Ideal energy absorber

The intrusion levels, developed analytically above, hold true when the contact area remains constant throughout the impact. In this case the stress-strain curve for the ideal absorber follows the form shown in figure 3. This is also shown numerically using computer-aided dynamic simulation of a flat "leg form" with a contact area as assumed above in equation (10), over the complete height of the leg, as shown in figure 8. The knee joint was omitted in this case to look purely at the effects on acceleration. The dynamics of the impact have an influence on the acceleration, but a relatively flat curve

can be achieved.

However, when the flat leg impacts an absorber system, which is wider than the leg itself, then shearing and edge effects may also influence the loading and the ideal behaviour is no longer valid (figure 9). The use of the same absorber behaviour then leads to a rise in the acceleration levels later in the compression phase.

Round rigid leg – Modified energy absorber

When the flat legform is replaced by a rigid, round legform with diameter d, which better represents the component used for testing (but still without a knee joint), then the contact area throughout the impact event, projected in the direction of impact, increases in a sinusoidal manner from 0 to $d \bullet h_l$. Accordingly, in order to maintain a uniform resisting force, the energy absorber should behave with a stress-strain behaviour as shown in figure 10. This is confirmed through numerical simulation using a rigid legform on an absorber stress-strain curve, modified to account for the changing area and realistic physical behaviour, as shown in figure 11.

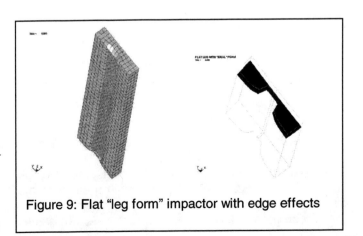

Figure 9: Flat "leg form" impactor with edge effects

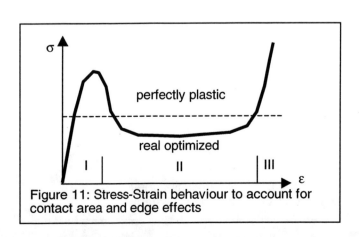

Figure 11: Stress-Strain behaviour to account for contact area and edge effects

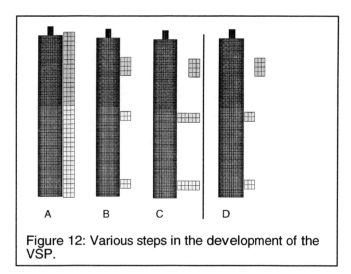

Figure 12: Various steps in the development of the VSP.

Round legform – with skin and foam – modified energy absorber behaviour and configuration

In further developing the understanding of the mechanisms at work for legform impact, the effects of adding the foam and skin, the knee joint and reducing the energy absorption regions to correspond to the main locations found in the real car fronts, the VSP is tuned to fit more practical situations. A series of studies showing the progression of the methodology is summarised in figure 12. The configuration shown in A represents a legform impact on an absorber acting over the complete height of the legform. B isolates the energy absorption regions to those better describing the areas in the car which can be used to take the impact. The initial analyses using these configurations were conducted using a knee joint where bending and shear were not allowed. This gave insight into the influence of the foam and skin of the legform components on the acceleration, since these also play a role in the energy absorption. These analyses were then conducted with the full legform as used in the tests, including the knee joint. By introducing the knee joint (figure 13), the relative distribution of the COGs of the femur and tibia must be

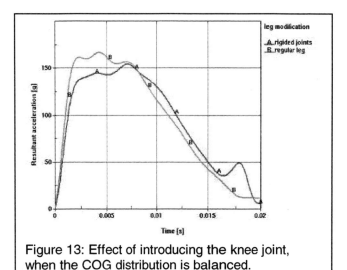

Figure 13: Effect of introducing the knee joint, when the COG distribution is balanced.

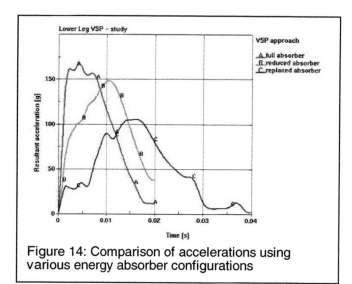

Figure 14: Comparison of accelerations using various energy absorber configurations

accounted for when distributing the stiffness of the absorbers, in order to limit the bending angle and the shear.

The set-up in C adjusts the geometry to model the offset between the different absorption areas, which normally do not act in the same plane. The energy absorber stiffness distribution and the positioning are set to maintain a bending angle of 0° and to avoid shearing at the knee. The resulting accelerations are shown in figure 14. The reduction in the maximum level of acceleration of C is mainly due to the increased packaging space used to maintain the bending angle at 0°. Further fine-tuning is normally required to fit the energy absorber behaviour to the design space available, as in D. This is very dependent on the geometry of the car front and the positions of components.

For acceleration curves shown in figure 14, the yield stress of the energy absorber is increased to account for the reduced impact area. As the set-up is adjusted to the front-end of the car and the design space available, the stiffness distribution of the VSP is modified. The initial stiffness (zone I in figure 11) accounts for the low initial contact area and resisting force. This transitions into a lower stress region (zone II) due to the stabilisation of the contact area. Finally, as a hard point is encountered, the system resistance increases (zone III). The goal is to absorb most of the energy before zone III is reached.

The VSP is further modified to account for the kinematics of the legform, allowing a certain amount of bending and shear, and converting some of the translational motion of the leg into rotation, hence reducing the amount of energy to be absorbed.

OPTIMISING THE REAL STIFFNESS PROFILE

The real stiffness profile can be measured through testing or numerical simulation. Whether measuring the profile experimentally or analytically, the part fixations, which can be assumed to be undeformable, will greatly

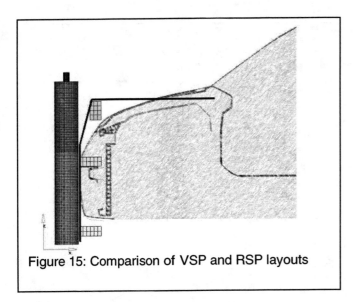

Figure 15: Comparison of VSP and RSP layouts

influence the stiffness behaviour. The generic VSP is compared to the actual profile of the target vehicle and the overlay shows where the vehicle profile needs to be stiffened or softened.

The first approach to optimising the real profile of the car is to adjust the stiffness distribution and force-intrusion behaviour, as shown in figure 11, compared to the VSP. The energy absorber system is modified as much as possible before any design changes to the outer shape of the car are considered.

When all stiffness modification options have been exhausted, the only remaining option is to recommend design (styling) changes to move the surfaces closer to the VSP.

EXAMPLE FOR LEGFORM IMPACT OPTIMISATION

The main emphasis in this paper has been on the legform impact situation. The overlay of a realistic car front end on a first representation of the VSP for that car is shown in figure 15. The VSP is first adjusted by moving the energy absorption region closer to those of the RSP. The stiffness distribution is then modified to account for the limited packaging space. In doing this the limits on acceleration, bending angle and shear specified by the car manufacturer as design limits are used to optimise the layout. Whenever possible some of the translational kinetic energy is converted into rotation of the legform to reduce the amount of energy to be absorbed in the car front. In many cases the solution is not physically possible within the space available and suggestions are made as to how the outer styling of the car could be modified to reach the specifications. This is done in close consultation with the car designers and engineers.

When the stiffness distribution and the final shape have been agreed upon using the VSP, the next step is to convert the design into something that can be achieved in reality. This means finding solutions to energy absorption requirements and tuning the absorber to fit the required behaviour.

ENERGY ABSORPTION SOLUTIONS

The current "traditional" bumper system normally uses a combination of an injection-moulded fascia with foam and possibly a second moulded part attached to the fascia or the beam. Such a system is already suitable for pedestrian safety, if the stiffness can be tuned to give the required response. The problem areas tend to be in the vicinity of fixation points and where light fittings are located. Figure 16 shows the response of a bumper system when loaded with a pedestrian leg at 11.1m/s and 13.4kg. The force level plateau is around 4kN before bottoming out after approximately 80mm. This is achieved through the use of a blow-moulded plastic part which fits between the bumper fascia and the beam and is designed to absorb energy. The load level was tuned by modifying the stiffness of the system using thickness and geometric shape.

For loading close to fixations and the edge of the bumper, the response is not as suitable. In this case the design of the fixations become very important. Here, a combination of a moulded plastic and a foam give best results. It is important when using foam systems that the efficiency is as high as possible. In order to minimise intrusion levels and maximise efficiency a foam system has been developed by Dow Automotive. This foam system, STRANDFOAM EA** PP oriented foams, displays a highly efficient load-intrusion response, when compared to other foam systems (see figure 17). The foam is based on strands of PP foam which are fused together to form a honeycomb system [5]. The force

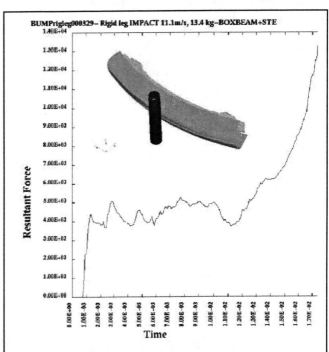

Figure 16: Force-Time response of a legform impact on a bumper using moulded plastic parts.

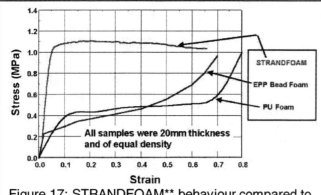

Figure 17: STRANDFOAM** behaviour compared to traditional foams. The stress level plateau can be tuned using density.

levels can be tuned based on density and geometric packing of the honeycombs. Using this foam, it is possible to maintain a broader range of options for the designer in terms of styling, minimising the packaging space required and tuning the response to fit the impact requirements.

The effect of using such an energy absorber is shown in figures 18 and 19. The amount of energy absorbed in the bumper foam is increased by a third and the acceleration levels are greatly reduced by tuning the stiffness distribution, building up the resisting force early in the impact event and removing energy before the hard point is encountered.

The energy absorption response of the total system is built up by superimposing the energy absorbed by individual components, such as the bumper fascia, followed by the bumper foam and lower stiffener and finally the hard points of the bumper beam. These individual contributions must be tuned to give an overall behaviour as defined by the VSP. This is demonstrated in figure 20. Initial contact with the bumper fascia must

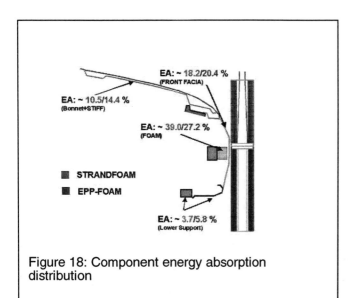

Figure 18: Component energy absorption distribution

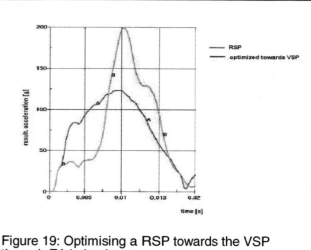

Figure 19: Optimising a RSP towards the VSP through EA behaviour

provide a relatively high stress on the legform, since this will act over the smallest contact area. This is achieved through the momentum effects of the mass of the fascia and by ensuring that the stiffness is high enough to compensate for the lower area. As the intrusion proceeds, the area of contact increases due to the deforming legform skin and foam and due to the round shape of the legform intruding into the bumper system. The stress levels should then decrease to maintain the acceleration at a constant level. This is achieved by a buckling effect in the bumper fascia, where the resisting force reduces dramatically, and the activation of the foam system to transition to a lower stress level response. The final bottoming-out effect against the bumper beam causes the stress level to rise at the end of the intrusion.

OTHER SYSTEM REQUIREMENTS

Once optimised for pedestrian impact, the system can then be checked against low speed (parking), insurance and high-speed impact conditions. Any modifications made to meet these tests should then be checked back against the VSP and secondary iterations run to establish the effect on pedestrian impact.

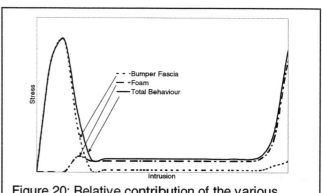

Figure 20: Relative contribution of the various components to energy absorption

CONCLUSIONS

Bumper design and engineering must balance knee acceleration (mainly influenced by stiffness) with knee bending and shear (influence more by geometric and stiffness profile than absolute stiffness). High efficiency foams such as STRANDFOAM EA** energy absorbing foams can control knee acceleration in the minimum packaging space, allowing the designer more freedom with regard to the front end profile.

The stiffness of the bonnet and the upper portion of the wings generally needs to be reduced and deflection under impact increased in order to maintain HPC<1000. The automotive industry is developing both active systems (deployable bonnets for example) and passive systems (such as foams or other energy absorbing structures) which can enable tuning of the impact response (energy absorption) as well as the stiffness of the bonnet and wings to withstand traditional loading.

The VSP represents the best case regarding the requirements for pedestrian safety. It can be used to identify weak points on the RSP and modify them through design and material decisions for the optimisation.

The ideal energy absorbing system behaves like rigid perfectly plastic materials. However, due to the changing contact area with the round legform, the total system response must be quite stiff initially and soften as the legform intrudes into the bumper. The initial high stiffness can be taken by the bumper fascia with the second phase of the energy absorption being taken by a more ideal elastic-plastic behaviour. For optimisation, materials, which represent such a behaviour, like STRANDFOAM EA**, are used to get as close as possible to the optimal energy absorption behaviour. A blow-moulded structure, for example, is also suitable to tune the stiffness behaviour to achieve such a response, as shown in figure 16.

Further publications will discuss the approach as it is applied to pedestrian head and upper leg impact, as well as the interaction between the various impact areas in the virtual stiffness profile.

REFERENCES

1. http://europa.eu.int/comm/enterprise/automotive/pagesbackground/pedestrianprotection/index.htm

2. COMMUNICATION FROM THE COMMISSION TO THE COUNCIL AND THE EUROPEAN PARLIAMENT Pedestrian protection: Commitment by the European automobile industry, COM (2001) 389(02), COM document 501PC0389 (en)

3. Naughton, P., "Pedestrian Safety - Materials versus Design", Madymo Users' Conference, Paris, France, May 23-24, 2000.

4. Naughton, P., Cate, P., *"An Approach to Front-End System Design for Pedestrian Safety"*, SAE 2001-01-0353 paper presented at SAE Int. Congress & Exposition, Detroit, MI, USA, March, 2001.

5. Gandhe, Gajanan V., Tusim, Martin H., "High Efficiency Energy Absorption Olefinic Foam", SAE Technical Paer Series 199-01-0296, International Congress and Exposition, Detroit, March 1-4, 1999.

DEFINITIONS, ACRONYMS, ABBREVIATIONS

BL	bumper lead
BLEH	bonnet leading edge height
BLERL	bonnet leading edge reference line
COG	centre of gravity
GRL	ground reference level
LBH	lower bumper height
LBRL	lower bumper reference line
RSP	real stiffness profile
UBH	upper bumper height
UBRL	upper bumper reference line
VSP	virtual stiffness profile

Improvement of Pedestrian Safety: Influence of Shape of Passenger Car-Front Structures Upon Pedestrian Kinematics and Injuries: Evaluation Based on 50 Cadaver Tests

C. Cavallero, D. Cesari, and M. Ramet
ONSER
Laboratoire des Chocs et de Biomécanique
Bron, France

P. Billault
P.S.A. SA·CITROËN · Centre Technique · VELIZY

J. Farisse, B. Seriat-Gautier, and J. Bonnoit
Faculté de Médecine · Laboratoire d'Anatomie · Marseille

ABSTRACT

Faced with the importance of road accidents involving pedestrians struck by passenger cars (18 % of those killed in road accidents in France), an experimental programme of vehicle/pedestrian impact analysis has been since 1979 developed. The programme is an example of teamwork between doctors and engineers.

The theme of this paper is to compare the influence of different vehicles used upon the consequences of impact, at a speed of 32 km/h and on the basis of tests with cadavers.

The results of this research show that there is a great similarity between the vehicles used. In spite of the differences of mass, of profile, of bonnet length, and of the position and the shape of the front bumper, the variations in terms of injury consequences, as well as the impact kinematics, are difficult to weigh up.

The improvement of pedestrian safety implies either changes affecting small areas and necessarily having limited effects, or alterations to the parts of the vehicles that are struck by pedestrians, such alterations involving characteristics that are specific to this type of protection.

THE EXPERIMENTAL STUDY which began in Marseilles in 1979 brings together as a team a car manufacturer (Peugeot-Citroen), a research institute (ONSER) and a laboratory of the Faculty of Medicine (Anatomy Laboratory). The experimental programme has made it possible to perform more than 150 tests either with dummies (the ONSER 50 dummy) or with cadavers (1).

The principal goals of this research are the reduction of injury consequences by the tunning of the front structures of the vehicle (in particular, concerning the contact between the lower limbs and the front scuttle) ; study of the influence on the overall injury assessment of changes affecting small areas ; the development of a mathematical model of the vehicle/pedestrian impact ; and the refinement of a dummy adapted to the problem of simulating pedestrian impacts.

During the course of the first phase of experiments, the conditions in which the impact reconstruction was carried out have been simplified so as to reduce the variation of certains parameters. The tests performed have thus been run with production models, the car being in a "dive" attitude throughout, with braking starting at impact. The subject is positioned along the median axis of the vehicle, is completely freed of support at the moment of impact, and is impacted according to two modes (frontally, at the right profile). Several impact speeds, lying between 10 km/h and 40 km/h, have been used in the tests. (2)

The theme of our work is to compare the experimental test results, based upon cadaver tests, and using different vehicles at the same speed.

THE TEST EQUIPMENT USED

In the two selected impact modes (frontally and in profile), at a speed of 32 km/h, we have therefore chosen to perform tests using fresh cadavers.

At this speed, injuries always occur but the degree of their severity enables us to envisage a means of increasing protection and so to minimise their consequences (3).

The following production models have been used :

- CITROEN 2 CV (2 CV)
- CITROEN VISA (VISA)
- CITROEN GSA (GSA)
- CITROEN BX (X/GS - BX)
- PEUGEOT 505 (P 505)
- PEUGEOT 505, modified scuttle (P 505 *)

All these vehicles are different in terms of their mass (from 635 to 1350 kg), their front profile, and their bonnet length, but also in terms of the technical solutions adopted for their front bumper development (design, position, structures). The main characteristics of the vehicles used in the tests appear in an Annex. (Tab. n°1 et n°2)

For each type of impact (vehicle type, impact mode), we have performed 4 successive tests, except in the case of the X/GS - BX with which vehicle 5 tests were run. The main characteristics of the cadavers used appear in an Annex. (Tab n° 3)

We observe that the average age of subjects is 72/73 years, whereas the range of ages runs from 46 to 92 years. The subjects are representative of an age group that has what accident analysis shows to be a high accident risk. (4)

The breakdown of subject's height (fig.n°1) shows the 50th percentile as being at 1.735 m and therefore near to the normal statistical distribution.

Average weight is 65.5 kg.

INJURY ASSESSMENTS

The evaluation of injury severity has been performed using "AIS 81" (5). We have complemented this evaluation (MAIS) by calculating the ISS value for each subject.

We set out the features of this analysis in an Annex. (Tab. n° 3) making it clear that the body segments to which reference is made are those described in the ISS nomenclature :

- A : head and neck
- B : face
- C : thorax
- D : abdomen (internal injuries)
- E : extremities
- F : skin injuries

We are able to state that few injuries are of a higher level of severity than AIS 4 (there is one AIS 5 injury) ; that the upper limbs are only slightly injured (few injuries and all of them at a level of AIS 1) ; and that, when the AIS scaling method is applied there is no difference between the overall severity of frontal impacts (mean of AIS : 3.2) and of those to the profile (mean of AIS : 3.2). (fig n° 2)

Fig. n° 1 : DISTRIBUTION OF SUBJECT'S HEIGHT

Fig. n° 2 : GLOBAL SEVERITY

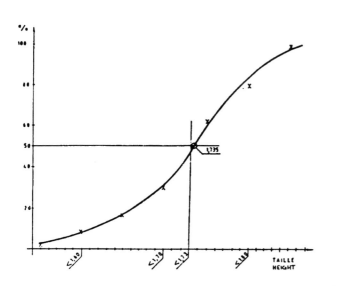

ALL VEHICLES	
	A.S. FACING
	MEAN OF ISS = 18.0 MEAN OF MAIS = 3.2
	A.S. IN PROFILE
	MEAN OF ISS = 16.9 MEAN OF MAIS = 3.2

Moreover, whatever the vehicle type and whatever the impact mode, it is the entire body that is involved in the distribution of injuries (Tab n° 4). The overall severity of anatomical subjects (MAIS) is not much different in all the cases, varying from 2.75 to 3.5 whereas the average ISS varies from 12.5 to 29 over a scale of severity ranging from 0 to 75.

We nevertheless observe two peculiarities in these results :

- A noteworthy variation exists between the "ISS" scores of certains vehicles as a function of the impact mode, except for the 2 CV and the GSA, the scores for these two cases being similar.

- Only the 2 CV is ranked at the lowest point in terms of ISS (mean of ISS :12.5) in both impact modes, and the same is true for the MAIS (mean of AIS : 2.75).

From the point of view of the injuries produced in each body segment (Tab n° 5), the frequency with which injuries are observed as well as the overall severity of the topographical region that is involved are similar for the vehicles as a whole. On the hand, a considerable improvement in one of the impact modes (frontal or profile) is often counter-balanced by a worsening in the other mode.

The ranking of injuries using the AIS scaling method does not make possible the distinguishing of different vehicles in terms of their relative aggressivness.

KINEMATICS

The analysis of films of the tests makes it possible to differentiate the two main stages of the impact sequence :

- The different impacts of the pedestrian against the vehicle (primary impact).
- The subject's impacts against the ground (secondary impact).

In the first impact stage, we have analysed two parts of the sequence in particular : the contact of the scuttle with the lower limbs, and the contact of the head of the subject with the vehicle. Concerning the second impact stage, we shall analyse the subject's head contact with the ground.

SCUTTLE/LOWER LIMB CONTACT - This is the injury-producing mechanism which is the characteristic one for this body segment. Analysis of the clinical results has allowed us to determine the main categories of injury observed, as well as the mechanisms by which they are produced (6). Ligament injuries are thus characteristic of frontal impacts whereas fractures are associated with impacts to the profile : these two injury types can, however, be found together in the same impact mode.

Fig. 3 shows the aratomical sites of the lower limb that are injured (AIS $\geqslant$ 2) as a function of vehicle type.

Fig. n° 3 : INJURIES (AIS   2) OF THE LOWER LIMB

	2 CV	VISA	GSA	BX ⊕	P 505	P 505 ✱
Femur		1		2 (2)		1
Knee	1	2	2	6 (5)	3	4
Ligaments of Knee	2	13	9	17 (14)	14	11
Tibia	4	4	4	5 (4)	2	3
Tibia/ Fibula	4	3	4		3	
Fibula	2	3	3	4 (4)	4	6
Ankle	7	1	2			2
Ligaments of Ankle	2		2			3
TOTAL	22	27	26	34 (30)	26	30

⊕ : 5 tests are realised with XB but only 4 with other cars

( ): 4/5 of all observed injuries

13

The influence of a low contact point (2 CV and P 505 *) shows up in the form of ankle injuries.

The limiting of the lower limb's angulation, however, between the ground, the scuttle and the sharp edge of the bonnet, performed with the P 505 * does not give rise to a gain in terms of knee injury. Only the 2 CV produces a lessening of this type of injury, this being due to its absence of a sharp bonnet edge and to the very low contact point.

The injuries at higher points of the lower limb (femur and femoral condyles) are characteristics of the scuttle's high position (X/GS-BX) or of the bonnet's sharp edge (P 505 and P 505 *).

With the exception of the 2 CV - whose front profile and bumper position and structure are very unusual - all the vehicles used behave in a similar way in spite of their diversity in terms of scuttle structure.

HEAD/VEHICLE CONTACT - Film analysis and the use of a graphic method (1) have enabled us to calculate the speed of the subject's head contact with the vehicle, as well as the part of the vehicle that was struck (Tab. n°6).

We have plotted on graphs (fig. n° 4 and n° 5) the values for the ratio of contact speed to initial vehicle speed (VTA/Vo). Average contact speed is higher when the subject is impacted frontally, except in the case of the 2 CV. In such cases of frontal impact, the average speed - VTA - is not much different from one vehicle to another. It is not influenced by the profile or by the mass of the vehicle and it is always above the vehicle's initial speed - Vo. When impact is to the profile, average speed - VTA - shows more scatter as a function of the vehicle type, but without there being any obvious correlation with the profile and/or the mass of the vehicle.

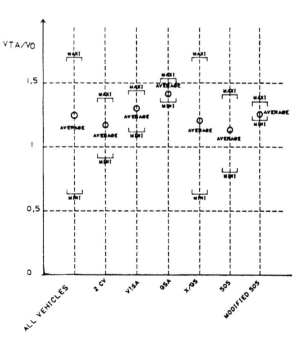

Fig N° 4. COMPARISON OF THE SPEED RATIOS VTA/Vo BY VEHICULE FOR VO = 32 KM/H (8.88 M/S)

A.S. FACING

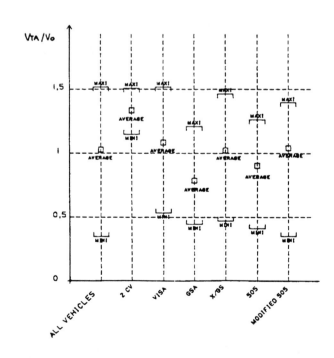

Fig N°5- COMPARISON OF THE SPEED RATIOS VTA/Vo BY VEHICULE FOR VO = 32 KM/H (8.88 M/S)

A.S. IN PROFILE

14

HEAD/GROUND CONTACT - The fall of the ground has seemed to us to be a random phenomenon because of the wide scatter of the attitudes taken up by pedestrians at ground contact, this being the case even where the vehicle, the impact mode and the impact speed are constant.

The calculation has been made of the average speed of the head in relation to the ground (Fig. n° 9). As a function of vehicle type or of impact mode there are no significant differences between these speeds.

Fig. n° 9 : AVERAGE SPEED OF THE HEAD IN RELATION TO THE GROUND

FACE

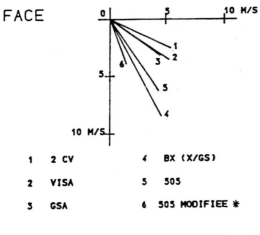

1	2 CV	4	BX (X/GS)
2	VISA	5	505
3	GSA	6	505 MODIFIEE *

PROFILE

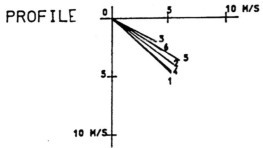

IT DOES NOT SEEM, from the results of our experiments, that a certain vehicle profile is more favourable to the pedestrian during the fall-to-ground stage of the overall impact sequence.

The comparison of the different profiles of production models of passenger car, in terms of evaluating the consequences of pedestrian impacts, is difficult to perform.

In the case of a uniform initial speed, and in spite of the differences of mass, of profile, of bonnet length, and of bumper position and shape, the analysis of the injury assessment and of the subject's kinematics, the ranking of the vehicles studied represents a difficult task.

The ranking of injuries using the AIS scaling system does not make it possible to characterise each one of the vehicles. This similarity of results may stem from :

- A deficiency of the AIS scaling system when evaluating the severity of experimental impacts against cadavers, due to the small number of injuries described and the impossibility of taking into account the association of injuries within the same body segment or the association of several injuries body segments. Additionally, the restricted ranking scale of the AIS system masks possible variations in severity.

- A too great similarity, in terms of their profile and/or their structures, between the vehicles studied.

- The production, during the fall-to-the-ground stage, of injuries that are sufficiently serious to hide - at the time when the AIS ranking is attributed - the specific injuries engendered during the primary impact.

- The influence of injuries that are specific to the use of cadavers and/or the absence of injuries that are not able to be reproduced by the type of test (head and intra-abdominal injuries).

The kinematic behaviours recorded have not shown significant variations from one vehicle to another, this similarity seeming to be the result of the likeness that exists between current passenger car models, a likeness that is conceptual and structural.

As for injuries to the lower limbs, the experiments have demonstrated that the injury risk is pratically identical, but that it is possible to change the location of the main injuries (case of the 2 CV). The objective of protecting front scuttles (low speed impacts) is difficultly compatible with that of protecting pedestrians because of the high stiffness and the unfavourable position of the feature that is impacted.

Analysis of the head/vehicle contact position (Fig. n° 6) shows a small variation between the different vehicles in terms of the distance between the lower point of the windscreen and the ground and of the intermediate area between the bonnet and the windscreen. The frequency of impacts into the windscreen is higher for the highest L values.

But this notion of the position of the impacted area of the vehicle (L) must be related to the subject's height (T), and we have thus calculated the L/T relationship (Tab n°7). We observe that the values are very close to one another, whatever the vehicle concerned. (Fig n°7 and n°8)

The impact speed at which the subject's head strikes the vehicle can not be considered as a term enabling us to differentiate between the influence of the profile of the various vehicles involved in our experiments. The head's contact point on the vehicle (if contact there is) is directly related to the ratio between the full height of the subject's profile (L/T), those vehicles having a small ground-to-windscreen distance (LPB) tending to lead to impacts against this vehicle feature.

Fig. n° 6 : IMPACTED AREA BY THE HEAD ON THE VEHICLE

	Windscreen	Cowl	Bonnet	No contact	L
2 CV	6	2			1 895
VISA	8				1 650
GSA	4	4			1 840
X/GS - BX	7	2		1	1 760
P 505	5	3			1 910
P 505 ✗	5	1	2		1 910

Fig. N° 7 _ COMPARISON OF THE LENGHT RATIOS L/T BY VEHICULE FOR VO = 32 KM/H (8.88 M/S)

A.S. FACING

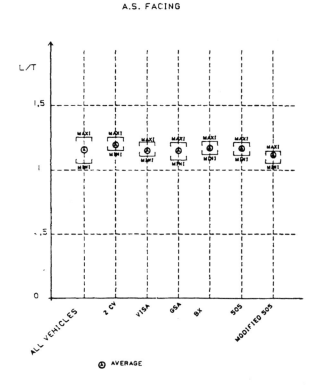

Ⓐ AVERAGE

Fig. N° 8 _ COMPARISON OF THE LENGHT RATIOS L/T BY VEHICULE FOR VO = 32 KM/H (8.88 M/S)

A.S. IN PROFILE

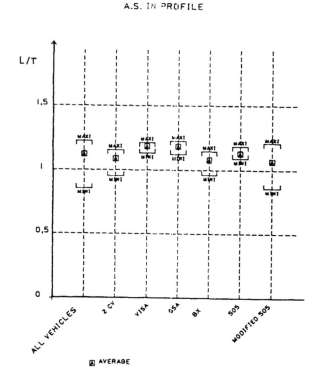

Ⓐ AVERAGE

Pedestrian head contact with the vehicle does not seem to be influenced by the vehicle's profile - in terms of impact speed - but the area of the vehicle that is struck is directly related to the shape of the profile as well as to the height of the subject. It is, therefore, possible to predict the probable impacted areas as a function of the full height of the subject's profile (L/T ratio).

The factors that are customarily compared (AIS - ISS - the subject's impact speed against the vehicle or the ground) lack position in this analysis, and it is only thanks to a more finely graduated research study, taking into account the topography of injuries their co-existence and their incapacitating consequences, the study also noting the precise way in which injuries are generated, that it will be possible to evaluate the influence of different car structures (their shape and their stiffness).

In point of fact, current passenger car models are similar to one another with regard to the injuries caused to pedestrians impacted in the middle of the spide range. The improvement of the safety of pedestrians thus implies the making of innovatory changes to vehicle front structures, and it necessitates either changes affecting small areas and necessarily having limited effects, or major alterations to the parts of the vehicle that are struck by pedestrians, taking into account the overall injury consequences of impact, such alterations involving characteristics that are specific to this type of protection.

REFERENCES

1 - P. BILLAULT - M. BERTHOMMIER : Centre
    Technique CITROEN
    J. GAMBARELLI - G. GUERINEL - J. FARISSE -
    B. SERIAT-GAUTIER - N. DAOU : Faculté de
    Médecine de MARSEILLE
    P. BOURRET - C. CAVALLERO - D. CFSARI -
    M. RAMET : ONSER

    "PEDESTRIAN SAFETY IMPROVEMENT RESEARCH"
    8th ESV - Wolfsburg - Oct. 1980

2 - P. BOURRET - J. FARISSE - B. SERIAT-GAUTIER:
    Faculté de Médecine de MARSEILLE
    R. LAROUSSE - P. BILLAULT : CITROEN, PARIS
    M. RAMET - D. CESARI - C. CAVALLERO : ONSER
    Laboratoire des Chocs, BRON

    "EXPERIMENTAL STUDY OF INJURIES OBSERVED ON
    PEDESTRIANS"
    IVth IRCOBI - Goeteborg - 1979

3 - D. CESARI - M. RAMET - C. CAVALLERO : ONSER
    BRON
    P. BILLAULT : CITROEN, VELIZY
    J. GAMBARELLI - G. GUERINEL - P. BOURRET -

J. FARISSE - B. SERIAT-GAUTIER : Faculté
de Médecine de MARSEILLE

    "EXPERIMENTAL STUDY OF PEDESTRIAN KINEMA-
    TICS AND INJURIES"
    Vth IRCOBI - Birmingham - 1980

4 - K. LANGWIEDER - M. DAUNER - W. WACHTER -
    TH. HUMMEL : German Association of Third
    party, Accident, Motor vehicle and Legal
    Protection Insurers (HUK-VERBAND) Automo-
    bile Engineering Department

    "PATTERNS OF MULTI-TRAUMATISATION IN PEDES-
    TRIAN ACCIDENTS IN RELATION TO INJURY COM-
    BINATION AND CAR SHAPE"
    8th ESV - Wolfsburg - Oct 1980

5 - THE ABREVIATED INJURY SCALE - 1980
    Revision A.A.A.M. 1980

6 - J. FARISSE - B. SERIAT-GAUTIER - J. DALMAS-
    N. DAOU : Faculté de Médecine de MARSEILLE
    P. BOURRET - C. CAVALLERO - D. CESARI -
    M. RAMET : ONSER, BRON
    P. BILLAULT - M. BERTHOMMIER : CITROEN,
    VELIZY

    "ANATOMICAL AND BIOMECHANICAL STUDY OF
    INJURIES OBSERVED DURING EXPERIMENTAL
    PEDESTRIAN-CAR COLLISIONS"
    VIth IRCOBI - Salon-de-Provence - 1981

APPENDIX

TAB. n° 1 and TAB. n° 2 : Profile of vehicles in breaking position.

TAB. n° 3 : Main characteristics of anatomical subjects.

TAB. n° 4 : Overall severity of anatomical subjects.

TAB. n° 5 : Observed injuries in relation to the severity.

TAB. n° 6 : Graphic method to determine the speed of the head, in relation to the vehicle.

TAB. n° 7 : Determination of the position of the impacted area.

TAB. n° 1 and TAB. n° 2 : PROFILE OF VEHICLES IN BREAKING POSITION

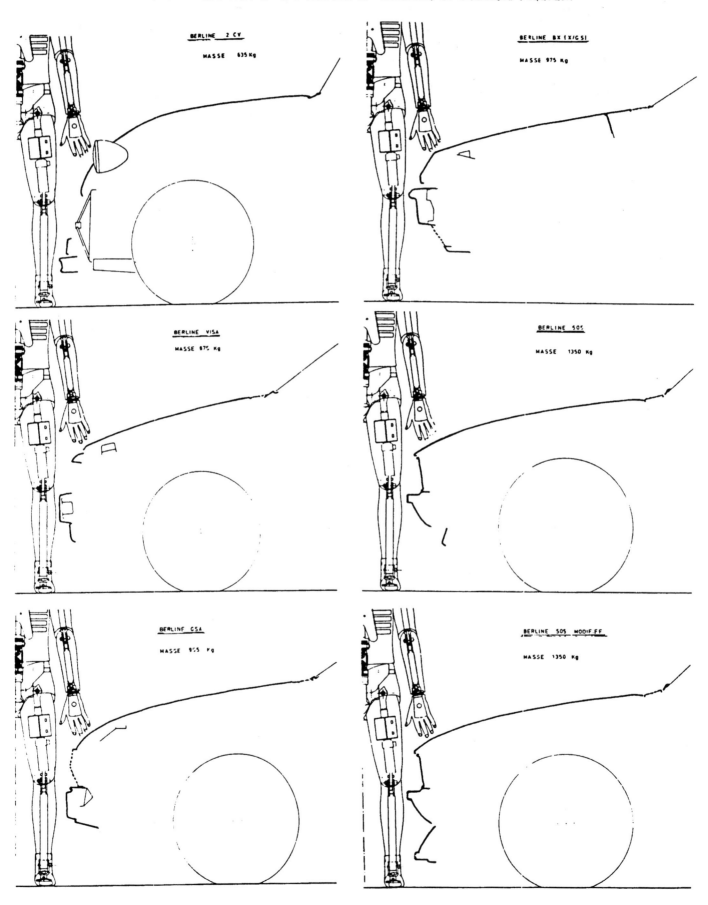

TAB. n° 4 : OVERALL SEVERITY OF ANATOMICAL SUBJECTS

VISA	GSA	BX	2 CV	P505	P505 *
M1 =22.25 M2 = 3.5	M1 =19 M2 = 3.5	M1 =29 M2 = 3.6	M1 =12.5 M2 = 2.75	M1 =13 M2 = 2.75	M1 =12.75 M2 = 3
M1 =17.25. M2 = 3	M1 =19.25 M2 = 3.5	M1 =15.2 M2 = 3.2	M1 =12.5 M2 = 2.75	M1 =17.25 M2 = 3.25	M1 =20.25 M2 = 3.5

TAB. n° 3 : MAIN CHARACTERISTICS OF ANATOMICAL SUBJECTS

PROFILE	FOC	S	A (years)	H (cm)	W (kg)	AIS - Segments A	B	C	D	E	F	MAIS	ISS
2 CV	97	♂	75	172	57					2	1	2	5
2 CV	101	♂	46	182	70	3		3		2	1	3	22
2 CV	102	♂	64	184	74			3		2	1	3	14
2 CV	103	♂	69	170	56					3	1	3	10
VISA	17	♂	69	160	54			3		3	1	3	19
VISA	39	♀	75	167	55			3		3	1	3	18
VISA	46	♂	70	184	74	3	2			3	1	3	22
VISA	62	♂	70	173	67					3	1	3	10
GSA	14	♂	77	176	53			1		3	1	3	11
GSA	55	♂	79	173	75			4		3	1	4	26
GSA	56	♀	75	168	81			3		3	1	3	19
GSA	91	♀	67	177	67	2		4		2	1	4	21
505	79	♂	65	178	69			3		3	1	3	19
505	84	♂	59	179	70					3	1	3	10
505	87	♂	67	175	70			4		3	1	4	26
505	89	♀	76	174	62	5				3	1	3	14
505 *	111	♀	87	159	55			1		3	1	3	11
505 *	113	♂	72	173	59					3	1	3	10
505 *	120	♂	56	173	64	3				3	1	3	10
505 *	121	♀	91	169	56				4	3	1	5	50
XB/GS	64	♀	76	157	56			1		3	1	3	11
XB/GS	65	♂	60	187	86					3	1	3	10
XB/GS	70	♂	75	181	85	3		4		3	1	4	34
XB/GS	71	♂	81	182	90			1		3	1	3	11
XB/GS	72	♂	60	169	69					3	1	3	10

FACE	FOC	S	A (years)	H (cm)	W (kg)	AIS - Segments A	B	C	D	E	F	MAIS	ISS
2 CV	96	♂	67	182	50	2		2		3	1	3	17
2 CV	100	♂	78	174	70	3		2		3	1	3	22
2 CV	104	♂	64	176	66					2	1	2	5
2 CV	105	♂	60	175	70		1			2	1	3	6
VISA	16	♂	56	170	52			3		3	1	3	10
VISA	18	♂	89	175	55		4	3		3	1	4	34
VISA	43	♂	71	169	62			4		3	1	4	26
VISA	61	♀	51	172	56	3		1		3	1	3	19
GSA	13	♂	66	171	64			1		3	1	3	10
GSA	15	♂	89	155	51		4			3	1	4	26
GSA	54	♂	80	176	60			1	4	3	1	4	26
GSA	91	♀	73	162	66	3	2	1		3	1	3	14
505	78	♂	78	181	74	3	1	1		3	1	3	19
505	81	♂	67	181	65		1	1		3	1	3	11
505	88	♂	69	165	64			1		3	1	3	11
505	90	♀	69	165	64			1		3	1	3	11
505 *	107	♂	67	184	77	2		2		3	1	3	17
505 *	108	♂	77	180	71					3	1	3	10
505 *	109	♂	78	165	51					3	1	3	10
505 *	110	♂	72	175	70			2		3	1	3	14
XB/GS	63	♂	75	173	61	3		1		3	1	3	19
XB/GS	66	♂	92	176	62	4	3	3		3	1	4	41
XB/GS	67	♂	76	173	86	4	2	1		3	1	4	29
XB/GS	68	♂	75	172	65	4	2			3	1	4	29
XB/GS	69	♂	85	176	70	3		3		3	1	3	27

TAB. n° 5 : OBSERVED INJURIES IN RELATION TO THE SEVERITY

PROFILE	AIS	2 CV	VISA	GSA	505	505*	XB(•)
GLOBAL (MAIS)	1						
	2	1					
	3	3	4	2	3	3	4 (3)
	4			2	1	1	1 (1)
	5						
HEAD/NECK	1	1	3		2	1	4 (3)
	2				1		
	3	1	1			1	1 (1)
	4						
	5						
FACE	1	2	3	1	4	1	3 (2)
	2		1			2	
	3						
	4						
	5						
THORAX	1	1	1	2		3	5 (4)
	2			1	1		1 (1)
	3	2	2	1	2		1 (1)
	4			2	1		
	5						
ABDOMEN	1		1				
	2		2		1	1	2 (2)
	3					1	
	4						
	5						
UPPER LIMB	1	2	1	3	2	2	4 (3)
	2						
	3						
	4						
	5						
LOWER LIMB	1	5	1	1	2	3	6 (5)
	2	11	7	8	2	9	6 (5)
	3	2	8	2	9	7	9 (7)
	4						
	5						

FACE	AIS	2 CV	VISA	GSA	505	505*	XB(•)
GLOBAL (MAIS)	1						
	2	2	2	2			2 (2)
	3	2	2	2	4	4	3 (2)
	4						
	5						
HEAD/NECK	1	2		1	3	1	2 (2)
	2	2					1 (1)
	3	1	1		1		4 (3)
	4						3 (2)
	5						
FACE	1	5	5	4	4	4	4 (3)
	2			3			2 (2)
	3						1 (1)
	4		1	1			
	5						
THORAX	1	2		2	6	1	4 (3)
	2	2				2	
	3		4				2 (2)
	4		1	1			1 (1)
	5						
ABDOMEN	1			1	1	1	
	2		1				
	3			1			
	4						
	5						
UPPER LIMB	1	3	1	1	1	1	2 (2)
	2						
	3						
	4						
	5						
LOWER LIMB	1	5	3	2	2	4	6 (5)
	2	5	3	4	6	5	13 (10)
	3	4	9	4	8	6	
	4						
	5						

TAB. n° 6 : GRAPHIC METHOD TO DETERMINE THE SPEED OF THE HEAD IN RELATION TO THE VEHICLE

ESSAI FOC 101

$V_0 = 9$ M/S

S.A.-- PROFIL -- 70KG

T_0	0	T_A	0,154
T_1	0,04		
T_2	0,09		

$VT_A = 13,57$ M/S $VT_{A(N)} = 13,02$ M/S

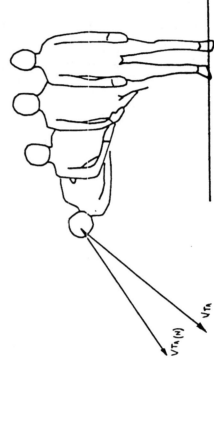

ESSAI FOC 100

$V_0 = 8,90$ M/S

S.A.-- FACE -- 70 KG

T_0	0	T_A	0,130
T_1	0,04		
T_2	0,09		

$VT_A = 12,31$ M/S $VT_{A(N)} = 12,04$ M/S

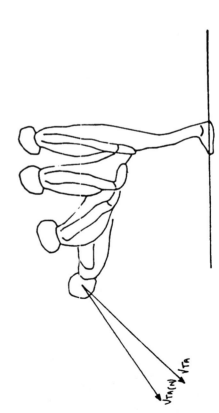

22

TAB. n° 7 : DETERMINATION OF THE POSITION OF
           THE IMPACTED AREA

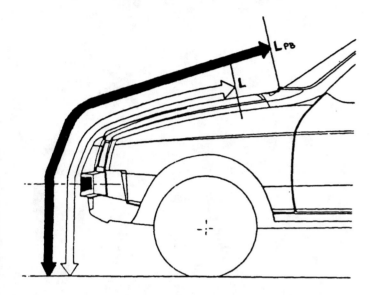

Engineering Thermoplastic Energy Absorber Solutions for Pedestrian Impact

Dominic McMahon, Frank Mooijman and Stephen Shuler
GE Plastics

ABSTRACT

This paper will describe an approach to satisfying proposed European Enhanced Vehicle Safety Committee (EEVC) legislation for lower leg pedestrian impact. The solution for lower leg protection is achieved through a combination of material properties and design. Using Computer Aided Engineering (CAE) modeling, the performance of an energy absorber (EA) concept was analyzed for knee bending angle, knee shear displacement, and tibia acceleration. The modeling approach presented here includes a sensitivity analysis to first identify key material and geometric parameters, followed by an optimization process to create a functional design.

Figure 1. PC/PBT injection molded Energy Absorber

Results demonstrate how an EA system designed with a polycarbonate/polybutyelene terephthalate (PC/PBT) resin blend, as illustrated in Figure 1, can meet proposed pedestrian safety requirements. Engineering thermoplastics offer higher energy absorption efficiency and a more consistent impact performance over a range of temperatures, than conventional foam systems. This permits the engineering of pedestrian safety solutions that fit within current styling and packaging space specifications.

INTRODUCTION

Pedestrian safety is now a high profile issue within the automotive industry. Accident investigations show that there are three areas of the pedestrian's body that are most subject to injury, and each of these is associated with an area of the car. The head is usually injured by the hood top, fender tops and A-pillars; the pelvis and upper leg by hood leading edge; and the knee and lower leg through contact with the bumper. The European Union and the Japanese government are both considering directives or guidelines to assess the risk to pedestrians from passenger cars during an accident. European New Car Assessment Programme (EuroNCAP) has an ongoing program to test and rate mainstream vehicles available on the European market for pedestrian safety. It uses an approach that is similar to the one being considered by the European Union. Poor results from tests conducted on most mainstream vehicles, which were not designed to be meet the tests clearly illustrate the need for new design ideas that will meet the requirements without adversely affecting any other performance requirements.

As illustrated in Figure 2, the current pedestrian safety assessment procedure consists of several different tests that represent the impact of the leg, upper leg and head. The lower leg test typically involves the front bumper and, to a lesser extent, the grille, hood and headlights. The front bumper is a key part of any discussion regarding lower leg impact, and – more specifically - the energy management capability of the bumper is the subject for this paper.

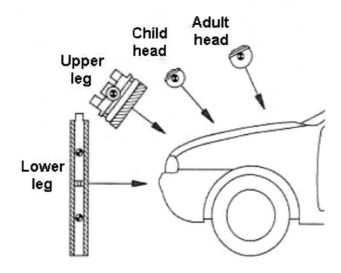

Figure 2: Pedestrian impact criteria tests

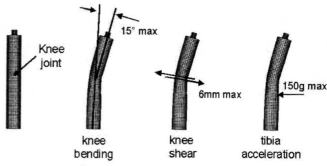

Figure 4: Leg impact criteria

LOWER LEG REQUIREMENTS

The lower leg test is carried out by impacting the leg-form into the front of the vehicle as pictured in Figure 3.

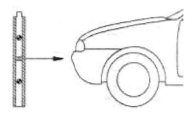

Lower Leg

Speed = 40kph
= 11.1m/s
Mass = 13.4 kg
Energy = 825 J

Figure 3: Lower leg impact test

Illustrated in Figure 4, the proposed legislation has three criteria to be met:

1. Angular rotation of the leg about the knee <15°
2. Lateral shear at the knee < 6 mm
3. Peak deceleration < 150g

A study of a number of different vehicles revealed that, frequently, the most difficult of the criteria to achieve is angular rotation. Typically the approach taken to achieve improved rotation is to consider both a stiffening of the lower section of the bumper and an improvement in the energy management capability of the upper section. This dual approach can achieve good results, but a lower stiffening system will be a weight penalty and may compromise the damageability performance of the lower part of the bumper. More importantly, it may require a vehicle styling change, which is generally undesirable. The goal is to develop a relatively simple solution that can achieve the required rotation without the need for a lower support, while maintaining the flexibility of the lower bumper edge and avoiding changes to the vehicle styling.

ENERGY MANAGEMENT

Many different systems provide energy management for vehicle bumper systems, and new concepts designed to meet the proposed lower leg test are being evaluated. They all share a primary requirement: the need to use the available packaging space in the most efficient manner, while meeting low cost, weight and ease of manufacture criteria. Figure 5 illustrates the commonly accepted guidelines for defining the efficiency of an energy management system.

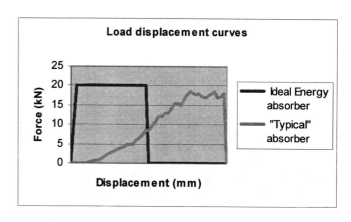

Figure 5: Load-displacement curves

The faster the response of the energy absorbing structure to the impact event, the more efficient the energy management and, therefore, the smaller the depth of space needed to absorb the energy from the event. A newly developed injection molded energy absorber made from a PC/PBT (XENOY® Resin) blend engineering thermoplastic is able to offer excellent energy management at a comparably low cost and weight, and can be particularly suitable for systems designed for lower leg performance.

THE NEW ENERGY ABSORBER

This PC/PBT blend, developed for exterior automotive applications, combines high modulus with high ductility, This engineering thermoplastic is able to deliver a steep reaction to an applied stress until it reaches the plastic deformation stress. Then it will deform plastically to produce a comparatively flat section of the stress strain/curve (Figure 6). The energy absorbed is the area under the curve. By developing an EA design that takes advantage of these properties, it is possible to bring a high level of energy management efficiency to the bumper system.

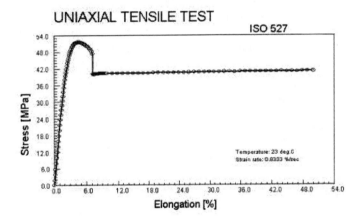

Figure 6: PC/PBT stress-strain curve

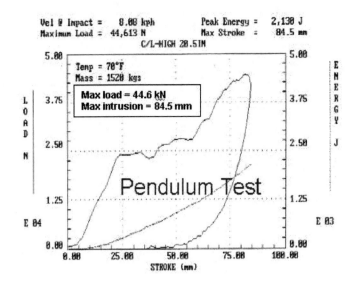

Figure 7: Pendulum impact test results

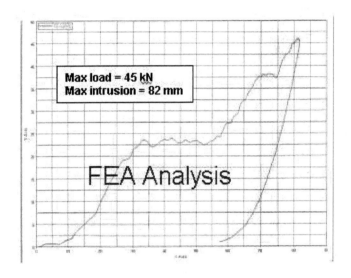

Figure 8: Finite Element Analysis of pendulum impact

TEST PROCEDURES

Finite element analysis (FEA) can have particular advantages over real life testing for lower leg impacts because the test itself is both time consuming and complex to set up [1,2,3,4]. The accuracy of the FEA will depend upon the material modeling of the leg form and the bumper/vehicle. As demonstrated in Figures 7 and 8, correlation of more than 95 percent has been achieved between FEA analysis and real life FMVSS tests by developing appropriate material models for the PC/PBT resin blend [6].

Using these results as a basis, a mainstream European production vehicle was selected to study lower leg performance and, in particular, to compare the capability of an injection molded PC/PBT absorber with an expanded polypropylene foam (EPP) absorber. EPP is widely used in bumper systems and it would be convenient if it could be simply adapted to provide lower leg performance.

Initially, we must consider typical load/displacement characteristics of EPP foam and PC/PBT energy absorbers. Figure 9 shows results from an 8 kph pendulum impact. The higher efficiency of the PC/PBT absorber allows the impact energy to be absorbed with less total intrusion.

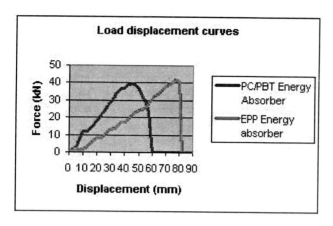

Figure 9: Pendulum impact data for EPP foam and PC/PBT energy absorbers

RESULTS

The EPP foam system of the case study vehicle was modified to optimize for lower leg impact performance. This involved lowering the density of the foam and filling all the available packaging space. The analysis showed that it is virtually impossible to comply with EEVC Working Group 17 (WG17) [5] directives when optimized foam is used on this vehicle (Table 1). However, an injection molded PC/PBT energy absorber inserted in the same available packaging space demonstrated that it is possible to meet, and exceed, the WG 17 directives, as shown in the analysis results depicted in Table 2. These results were obtained without changes to the styling of the vehicle.

EPP Foam Absorber:	
Peak Tibia Acceleration	164.7 g
Peak Knee Displacement	4.3 mm
Peak Knee Rotation	23.3°

Table 1: Lower leg impact results using EPP foam EA

PC/PBT Energy Absorber:	
Peak Tibia Acceleration	135.6 g
Peak Knee Displacement	2.6 mm
Peak Knee Rotation	14.4°

Table 2: Lower leg impact results using PC/PBT EA

Subsequent studies on various vehicle geometries have shown consistently good performance of the PC/PBT energy absorber during lower leg impacts.

A significant factor is the available packaging space in relation to the offset and stiffness of the lower spoiler. Studies have demonstrated that, with >70 mm packaging space, the PC/PBT energy absorber is capable of meeting the WG 17 limits with no additional lower edge support. When the packaging space is between 40 and 70 mm, the help of a lower edge stiffener may be required to meet the rotation criteria.

CONCLUSION

The EA concept outlined in this paper offers a glimpse at an engineering thermoplastic breakthrough that is expected to help automotive manufacturers and tiers develop front-end safety systems to meet proposed European Union pedestrian protection legislation. Material data and FEA indicate that this system should be able to meet the proposed system requirements without significant vehicle styling changes.

With further material and design development, it is anticipated that several comprehensive material and design ideas that are currently being explored will help automotive manufacturers develop solutions that can meet the additional requirements of the proposed regulations including those for hood top, fender top and leading edge.

REFERENCES

1. M Howard, A Thomas, W Kock, J Watson, R Hardy "Validation and Application of a Finite Element Pedestrian Humanoid Model for use in Pedestrian Accident Simulations".
2. I Kalliske, F Friesen "Improvements to Pedestrian Protection as Exemplified on a Standard Sized Car"
3. KC Clemo "The Practicalities of Pedestrian Protection"
4. P Schuster, B Staines "Determination of Bumper Styling and Engineering Parameters to Reduce Pedestrian Leg Injuries"
5. "Improved Test Methods to Evaluate Pedestrian Protection Afforded by Passenger Cars" 1998 EEVC Working Group 17 report
6. Reference work by D. Evans, S. Shuler, S. Santhanam, "Predicting the Bumper System Response of Engineering Thermoplastic Energy Absorbers with Steel Beams," SAE paper 2002-01-1228

DEFINITIONS, ACRONYMS, ABBREVIATIONS

EA: Energy Absorber

CAE: Computer Aided Engineering

FEA: Finite element Analysis

PC/PBT: Polycarbonate/Polybutylene Terephthalate

EPP: Expanded Polypropylene

ETP: Engineering Thermoplastic

XENOY is a registered trademark of General Electric Corporation.

2002-01-0019

Pedestrian Throw Kinematics in Forward Projection Collisions

Thomas F. Fugger, Jr., Bryan C. Randles and Jesse L. Wobrock
Accident Research and Biomechanics, Inc.

Jerry J. Eubanks
Automobile Collision Cause Analysis

ABSTRACT

Pedestrian crash kinematics have been well documented for automobile versus pedestrian collisions. However, there is not significant amount of data concerning impact of pedestrians with a high profile vehicle. A series of pedestrian crash tests using full-sized vans was performed to add to the existing database of forward projection pedestrian collisions and to compare the crash test data to existing forward throw equations. The aim of this study was to examine the trajectory behavior of the pedestrian in a forward projection impact and the effect of different friction-value surfaces when applying a pedestrian model to the data.

In performing the tests, the pedestrian dummy was stabilized using an 18.2 kg tensile strength monofilament wire hanging from a cantilever beam. The impacting vans were instrumented with a triaxial accelerometer triggered at impact with the dummy. Several testing surfaces were used, ranging from dry asphalt to a skidpad with > 1/16th inch depth of water. The trials were videotaped using two high-speed digital camcorders that recorded at 120 Hz and 240 Hz., respectively. Each trial was computer digitized and analyzed using motion analysis software. The results of the digitized analysis were compared to the measured throw distance and velocity profile integrated from the from the accelerometer data of the van. The impact speed versus throw distance was plotted and compared to existing forward projection throw models. Additionally, the trajectory of the dummy was analyzed for the forward projection collisions and compared to existing models and previous wrap trajectory data. Comparison of the pedestrian throw kinematics and vehicle kinematics were compared over the various testing surfaces.

INTRODUCTION

Pedestrian fatalities and injuries in traffic collisions are a significant problem. In 2000, 4739 pedestrians were killed and 78,000 were injured in automobile versus pedestrian collisions in the United States. On average, a pedestrian is killed in a traffic collision every 111 minutes and injured every 7 minutes (15). To reduce and reconstruct automobile versus pedestrian collisions, an understanding of vehicle-pedestrian interaction during the collision phase and post-impact pedestrian kinematics is needed. Pedestrian trajectories following vehicle impacts typically fall into one of six categories including wrap trajectories, fender vaults, roof vaults, somersaults, restricted fender vaults, and forward projections (6,16). This paper is concerned with forward projections, which are the second most common type of pedestrian trajectory (16).

A review of the scientific literature (1-14,16-23) revealed a considerable amount of published scientific data correlating pedestrian throw distance to vehicle impact speed for low profile vehicles (e.g. passenger cars). However, there is a limited amount of data available concerning high profile automobiles (e.g. cargo/passenger vans, SUVs, large trucks, etc.) versus pedestrian collisions. An impact between a high profile vehicle and a pedestrian typically results in a forward projection pedestrian trajectory. Forward projections occur because the impacting vehicle's vertical distance from the ground to the beginning of the hood (i.e. leading edge) is above the impacted pedestrian's center of gravity (CG). The impacted pedestrian is thrown forward as the pedestrian's CG does not move rearward relative to the vehicle's leading edge. (9). Although the pedestrian's head and upper torso may wrap rearward onto the vehicle hood and/or windshield, the pedestrian trajectory is still classified as a forward projection because the pedestrian's CG remains ahead of the vehicle.

Many variables need to be considered when investigating and reconstructing automobile versus pedestrian collisions (6,7). The two most prevalent variables currently used in accident reconstruction are pedestrian throw distance (measured from impact to rest) and vehicle impact velocity. Using a pedestrian throw equation, vehicular impact velocity can be estimated from a measured pedestrian throw distance. Evidence utilized to reconstruct automobile versus pedestrian collisions can include the point of impact, point of rest of objects, vehicle impact velocity, pedestrian throw distance, and post-impact pedestrian kinematics. Some other vehicle and pedestrian characteristics that can be considered include pre/post-impact vehicle braking, roadway surface (i.e. dirt, grass, asphalt, etc.), coefficient of friction (i.e. icy, wet, or dry roadway surfaces), traffic signal control timing, pedestrian gait velocity, intersection characteristics, and direction of pedestrian/vehicle travel, among others (6). The most important variable that needs to be identified when investigating an accident is vehicle impact velocity because it is correlated with pedestrian injury and fatality potential (4).

This study focuses on the interaction of high profile vehicles (i.e. cargo and passenger vans) and pedestrians during the collision phase and the post-impact pedestrian kinematics on two different surfaces. In a series of staged collisions, a pedestrian crash test dummy was impacted by a number of high profile vehicles on both wet and dry asphalt roadway surfaces. The purpose of the study was to add to the existing database of forward projection pedestrian collisions and describe the many variables involved in this complicated situation on both wet and dry roadway surfaces.

METHODOLOGY

PEDESTRIAN DUMMY - An Alderson Research Labs, Inc., Model CG-95 dummy was utilized in all of the pedestrian crash tests. This male dummy is a multi-segment model with articulating upper and lower extremities. The dummy stands approximately 1.8 m (75.5 inches) tall and weighs 76.8 kg (169 lbs). The dummy center of gravity location was in a location that equated to 47% of its height. For each of the tests, the dummy's clothing consisted of a wetsuit with coveralls on top and standard athletic footwear on the feet. Oil-based party makeup was applied to the rear of the dummy's head so that head strike with the vehicle front end could be easily identified.

TEST VEHICLES – Six vans were utilized in the testing program and the test series each was used for are listed in Table 1. The leading edge height for each van was above the CG of the test dummy. There was some slight variation in the amount of "hood distance" or the distance from the leading edge of the vehicle to the base of the windshield. The side profile of five of the test vans is shown in Figure 1. The profile of the Plymouth D100 was not available. As impact speeds increased and damage started occurring on the vans, the relative position of the dummy was moved to an "undamaged" portion of the van. This was done to minimize the potential for vehicle damage to affect the pedestrian kinematics. Each of the vans was instrumented with a triaxial block of accelerometers (IC Sensors 3031-050) that was affixed to the target vehicle's approximate static center of gravity. The acceleration data were used to determine the kinematic response of the center of gravity of the impacting vehicle.

Vehicle	VIN	Tests
1976 Ford Econoline 250	E25HHB68036	Dry
1971 Dodge B200	B22AB1V331738	Wet <1/16th
1982 Dodge B250 LWB	2B7HB23E8CU181643	Wet >1/16th
1980 Plymouth D100	BC2JTAX138862	Dry
1982 Chevy G20	1GCEG25FOC711203	Wet II
1977 Dodge Sportman	B21AB4X038813	Wet II

Table 1: Test Vehicles

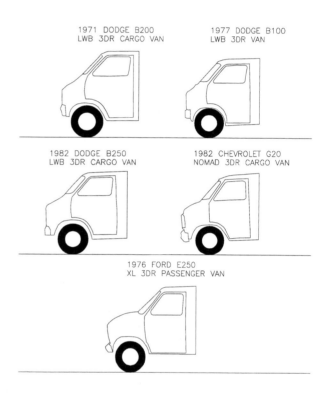

Figure 1: Side Profile of Test Vans

DATA ACQUISITION AND POST PROCESSING – All data were collected following the general theory of SAE Recommended Practice: Instrumentation for Impact Test - J211/1 Mar95 [23]. All accelerometer data were collected at 2000 Hz. Vehicle changes in velocity and displacements were calculated from vehicle acceleration data filtered with an SAE Class 180 filter.

TEST PROCEDURE - The pedestrian dummy was stabilized from a cantilever beam, for the dry surface tests, and a "lift" bucket for the wet surface tests. The dummy was standing on the ground with an 18.2 kg tensile strength monofilament line, used when necessary, between the dummy's head and the support device. This line was tightened to near the breaking point and provided a short period of lateral support for the dummy so that it would remain erect until impact. The line broke with minimal impact force and had no influence on the kinematics of the dummy based upon review of tests wherein the filament was used versus not used.

The trials were videotaped using two high-speed digital camcorders (JVC DVL 9500, JVC DVL 9800) that recorded at 120 Hz and 240 Hz, respectively. Each trial was computer digitized and analyzed using motion analysis software (Peak Motus™ 2000, Peak Performance Technologies, Inc., Englewood, CO). The results of the digitized analysis were compared to the measured throw distance and velocity profile integrated from the from the accelerometer data of the van. Fifty-six tests on dry asphalt pavement and eighty-four tests on wet asphalt pavement were performed using five different vans. The impact speeds ranged from 4 km/h to 60 km/h, with most of the data falling under 32 km/h. The impact speed versus throw distance was plotted and compared to existing forward-projection throw models.

The Statistical Package for the Social Sciences (SPSS®) was utilized for statistical analysis of the data, which consisted of regression analyses. Multiple forms of regression equations were calculated and the form that fit the data most reliably (highest consistent R^2 value) was reported in this paper.

RESULTS

FORWARD PROJECTION TRAJECTORY – DRY SURFACE - The trajectory of the pedestrian dummy, at low speed (4.42 kph) and a higher speed (34.5 kph) is shown in Figure 2 and Figure 3, respectively. As seen in these figures, the dummy does not appreciably rise into the air when struck by the van. An analysis of the trajectory for the dummy during the dry tests indicated an average takeoff angle of 1.78° with a standard deviation of ±2.1°. This is consistent with the expectations of a forward projection type of collision.

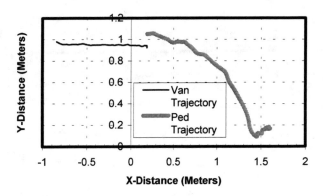

Figure 2: Forward projection trajectory for 4.42 kph impact on dry pavement

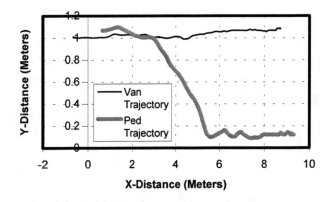

Figure 3: Forward projection trajectory for 34.5 kph impact on dry pavement

FORWARD PROJECTION TRAJECTORY – WET SURFACE - The trajectory of the pedestrian dummy, at low speed (4.42 kph) and a higher speed (25.77 kph) is shown in Figure 4 and Figure 5, respectively. As seen in these figures, the digitized dummy trajectory appears to rise slightly at impact, but within a short time, begins to head downward to the ground in a projectile type manner. An analysis of the trajectory for the dummy during the wet tests indicated an average takeoff angle of 2.6° with a standard deviation of ±2.7°. This is very similar to the results obtained for the dry tests.

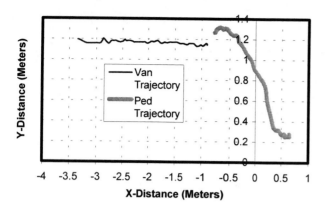

Figure 4: Forward projection trajectory for 4.42 kph impact on wet pavement

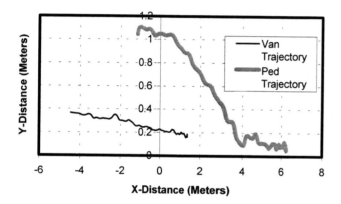

Figure 5: Forward projection trajectory for 25.7 kph impact on wet pavement

FORWARD PROJECTION THROW DISTANCE - The graph seen in Figure 6 represents the vehicle impact speed versus the pedestrian throw distance for all of the forward projection tests conducted in this current study. As seen in this figure, the data do not converge toward the origin. In these tests, it was observed, with the pedestrian dummy, that even for low impact speeds, the CG of the dummy fell forward away from the front of the van. Also seen in this figure is that the total distance the pedestrian dummy was displaced was greater, at a given impact speed, when the surface was wet.

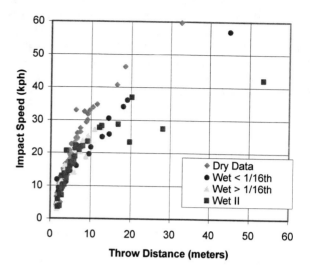

Figure 6: Vehicle impact speed versus pedestrian throw distance for forward projection tests.

Many different forward projection accident reconstruction equations were analyzed and compared to the gathered test data. Wood's "new" equations for forward projections (6) compared reasonably well to the test data.

Wood's equations for forward projections:

Wood – Low Value
$$S = 8.77 \times \sqrt{d_t}$$

Wood – High Value
$$S = 13.76 \times \sqrt{d_t}$$

d_t = throw distance in meters
S = vehicle speed in kilometers per hour

The measured test data compared to Wood's equations revealed a slight discrepancy. The forms of the equations require that the origin begin at (0,0). As seen in Figure 6, the data is actually shifted a bit to the right. Therefore, the forward projection equations were adjusted so that the origin began at the point where the throw distance was equal to the height of the pedestrian's CG. Figure 7 shows that after this adjustment was made the dry data fit reasonably well within the adjusted low to high range.

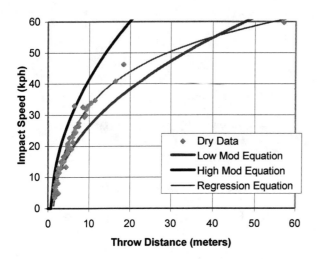

Figure 7: Impact speed versus throw distance, modified equations and regression analysis for dry data

Pedestrian dummy average deceleration during the slide/tumble phase was calculated by using the Peak motion analysis. The average deceleration values for the pedestrian dummy over the dry and wet asphalt testing surfaces are depicted in table 2. The average deceleration values were calculated from when the pedestrian dummy's CG y-velocity (downward) component equaled zero until the pedestrian's x-velocity (forward) component equaled zero.

Test #	Surface	Average Pedestrian Dummy Deceleration Value (Gs)
55-103, 149-156	Dry Asphalt	.43
104-125	Wet Asphalt, < 1/16th inch deep	.31
126-148	Wet Asphalt, > 1/16th inch deep	.37
157-196	Wet Asphalt II	.41

Table 2: Pedestrian dummy deceleration values for different testing surfaces

To account for the decreased coefficient of friction on the wet surfaces, Wood's equation was modified slightly again by multiplying the whole equation by a factor of the calculated deceleration value. The modified

equations as well as the regression equations are listed following figures 8, 9, and 10, which depict vehicle impact speed versus pedestrian throw distance for the three different wet asphalt surfaces. The natural logarithmic regression equations were found to be the most reliable (highest consistent R^2 value). It is noted that the origins of the regression equations do not begin at 0 meters, 0 kph.

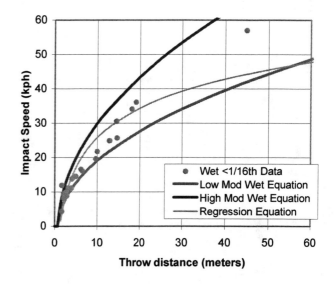

Figure 8: Impact speed versus throw distance, modified equations and regression analysis for wet < 1/16th inch data

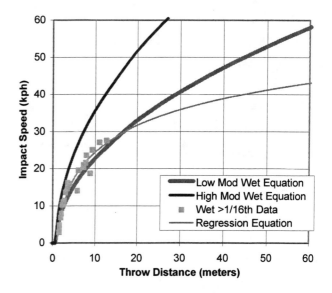

Figure 9: Impact speed versus throw distance, modified equations and regression analysis for wet > 1/16th inch data

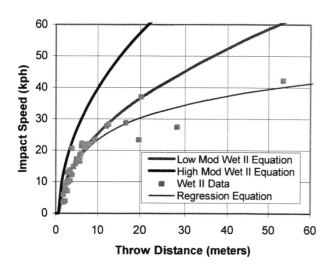

Figure 10: Impact speed versus throw distance, modified equations and regression analysis for wet II data

Regression equations:

Dry: $S = -4.0894 + (16.0339 \times \text{LN } d_t)$

$R^2 = .95$ Origin: (1.29 m, 0 kph)

Wet, < 1/16th: $S = -2.2350 + (12.2112 \times \text{LN } d_t)$

$R^2 = .88$ Origin: (1.20 m, 0 kph)

Wet, > 1/16th: $S = .6421 + (10.3645 \times \text{LN } d_t)$

$R^2 = .94$ Origin: (.94 m, 0 kph)

Wet II: $S = .2113 + (10.0365 \times \text{LN } d_t)$

$R^2 = .88$ Origin: (.98 m, 0 kph)

d_t = throw distance in meters
S = vehicle speed in kilometers per hour

Adjusted throw equations:

Dry Low Value:
$$S = 8.77 \times \sqrt{d_t - CG}$$

Dry High Value:
$$S = 13.76 \times \sqrt{d_t - CG}$$

Wet < 1/16th Low Value:
$$S = (8.77 \times \sqrt{d_t - CG}) \times (.31 / .43)$$

Wet < 1/16th High Value:
$$S = (13.76 \times \sqrt{d_t - CG}) \times (.31 / .43)$$

Wet > 1/16th Low Value:
$$S = (8.77 \times \sqrt{d_t - CG}) \times (.37 / .43)$$

Wet > 1/16th High Value:
$$S = (13.76 \times \sqrt{d_t - CG}) \times (.37 / .43)$$

Wet II Low Value:
$$S = (8.77 \times \sqrt{d_t - CG}) \times (.41 / .43)$$

Wet II High Value:
$$S = (13.76 \times \sqrt{d_t - CG}) \times (.41 / .43)$$

d_t = throw distance in meters
S = vehicle speed in kilometers per hour

DISCUSSION

Forward projection pedestrian impacts were conducted on both wet and dry pavement. Data obtained from these tests included the pedestrian post-impact trajectory, takeoff angle, slide/tumble deceleration rate and the overall throw distance. These tests show that the pedestrian dummy trajectory motion is not significantly different for the dry versus wet condition. A small variation in average takeoff angle was observed in the dry versus wet tests. The takeoff angle was evaluated against other parameters such as vehicle impact speed and van type but no correlation was discerned. The maximum takeoff angle observed in any individual test was around 14° in a wet surface test. This result was not typical as indicated by the average values of 1.78° and 2.6° obtained for the dry and wet surfaces, respectively.

However, the overall pedestrian movement (or throw distance) is a function of the pavement condition. This effect is small at low speeds and becomes readily apparent at speeds in excess of 20 kph. As one might expect, the pedestrian dummy was displaced further for a given impact speed on the wet surface when compared to a dry surface. This is consistent with the variation in average pedestrian slide/tumble deceleration rate observed in dry versus wet tests.

The fall and slide equations of motion were evaluated for modeling the experimental test data obtained from this series of tests. Initial evaluation of these equations did not provide adequate correlation and further investigation of the effect of takeoff angle and friction factor is required. Therefore, a regression analysis was performed. The regression equations obtained have good correlation to the data (R^2 of .88 to .95). The Wood "new" equations for forward projection pedestrian impacts were used to "bracket" the data. In the original form, the Wood equations did not correlate well with the data. This was due, primarily, to the offset in the low speed throw distance. In the subject crash tests, even at very low speeds, the dummy tended to fall over and land at least its CG distance away from the point of impact. Whether this accurately correlates to impacts involving humans is unknown at the present time. Utilization of the regression equations, at impact speeds under about 5 kph, should be done with close consideration of the physical evidence. Another modification to the Wood equation involved an adjustment for the average deceleration values obtained for the different surfaces. This methodology yielded upper and lower bound curves that, in all but the Wet Series II tests, resulted in a reasonable bound on the experimental test data.

CONCLUSION

One hundred and forty one forward projection pedestrian crash tests were conducted. The tests were conducted on dry and wet surfaces. These tests indicate that the pedestrian trajectory profile is similar on the dry versus wet surface. A small variation in pedestrian takeoff angle was observed on wet surfaces versus dry. Evaluation of the pedestrian deceleration during the slide/tumble phase of the trajectory showed a slight decrease in the average deceleration on the wet surface compared to the dry (.31 g's to .41 g's versus .43 g's). This is consistent with the greater throw distance observed in the wet surface tests.

A set of regression equations was developed for the experimental test data. The equations were found to correlate well with the experimental data. Upper and lower bounds for the data were found to be reasonably well bounded by slightly modified Wood forward projection equations.

Future work will examine very low friction surfaces (e.g. ice) and development of a methodology for testing a surface to incorporate a specific friction value or factor into a forward projection models.

ACKNOWLEDGMENTS

The authors express their appreciation to Jason Reyes, Doug Wilson, Kevin Cassidy and the California Association of Accident Reconstruction Specialists for their assistance in conducting the pedestrian crash tests. Appreciation is also extended to the Concord, California police department for the opportunity to use their vehicle test track for the wet surface pedestrian crash tests.

REFERENCES

1. Appel, H., Sturtz, G., & Gotzen, L. Influence of speed of collision and vehicular parameters on the severity of injury in pedestrian accidents. *Society of Automotive Engineers.* Paper No. 765038, 1976.
2. Aronberg, R. Airborne trajectory analysis derivation for use in accident reconstruction. *Society of Automotive Engineers.* Paper No. 900367, 1990.
3. Daniel, S. The role of the vehicle front end in pedestrian impact protection. *Society of Automotive Engineers.* Paper No. 820246, 1982.
4. Davis, GA. A simple threshold model relating pedestrian injury severity to impact speed in vehicle/pedestrian crashes. *Transportation Research Board.* Paper No. 01-0495, (in press).
5. Eubanks, J, & Haight, W. Pedestrian involved traffic collision reconstruction methodology. *Society of Automotive Engineers.* Paper No. 921591, 1992.
6. Eubanks, J. *Pedestrian Accident Reconstruction.* Lawyers & Judges Publishing Co., Tucson, Arizona, 1994.
7. Fricke, LB. *Traffic Accident Reconstruction.* Northwestern Traffic Institute, 1990.
8. Fugger, T, Randles, B, & Eubanks, J. Comparison of pedestrian accident reconstruction models to experimental test data for wrap trajectories. *ImechE Conference Transactions.* Paper No. C56/031/2000, 2000.
9. Happer, A, Araszewski, M, Toor, A, Overgaard, R, & Johal, R. Comprehensive analysis method for vehicle/pedestrian collisions. *Society of Automotive Engineers.* Paper No. 2000-01-0846, 2000.
10. Janssen, EG, & Wismans, J. Experimental and mathematical simulation of pedestrian-vehicle and cyclist-vehicle accidents. *Society of Automotive Engineers.* Paper No. 856113, 1985.
11. Kuehnel, A. Vehicle-pedestrian collision experiments with the use of a moving dummy. *Society of Automotive Engineers.* Paper No. 1974-12-0018, 1974.
12. Kuehnel, A, & Appel, H. First step to a pedestrian safety car. *Society of Automotive Engineers.* Paper No. 780901. 22nd Stapp Car Crash Conference, 1978.
13. Lucchini, E, & Weissner, R. Influence of bumper Adjustment on the kinematics of an impacted pedestrian. *International Research Council On Biomechanics of Impact*, 1978.
14. Lucchini, E, & Weissner, R. Differences between the kinematics and loading of impacted adults and children. Results from dummy tests. *International Research Council On Biomechanics of Impact*, 1980.
15. National Highway Traffic Safety Administration. *Traffic Safety Facts 2000 – Pedestrians.* National

Center for Statistics and Analysis, Washington D.C., pp. 1-6, 2000.

16. Ravani, B, Brougham, D., & Mason, RT. Pedestrian post-impact kinematics and injury patterns. *Society of Automotive Engineers.* Paper No. 811024, 1981.

17. Schneider, H, & Baier, G. Experiment and accident: comparison of dummy tests and real pedestrian accidents, the effects of vehicle frontal design on pedestrian injury in real accidents. *Society of Automotive Engineers.* Paper No. 1991-15-0026, 1991.

18. Searle, JA. The physics of throw distance is accident reconstruction. *Society of Automotive Engineers.* Paper No. 930659, 1993.

19. Severy, D, & Brink, H. Auto-pedestrian collision experiments. *Society of Automotive Engineers.* Paper No. 660080, 1966.

20. Stcherbatcheff, G, Tarriere, P, Duclos, P, Fayon, A, Got, C, and Patel, A. Simulation of collisions between pedestrians and vehicles using adult and child dummies. *Society of Automotive Engineers.* Paper No. 751167, 1975.

21. Sturtz, G, & Suren, E. Kinematic of real pedestrian and two-wheel rider accidents and special aspects of the pedestrian accident. *International Research Council On Biomechanics of Impact*, 1976.

22. Wood, D. Impact and movement of pedestrians in frontal collisions with vehicles. *ImechE Conference Transactions.* Paper No. 51/88, 1988.

23. Wood, D. Application of a pedestrian model to the determination of impact speed. *Society of Automotive Engineers.* Paper No. 910814, 1991.

CONTACT

Thomas F. Fugger, Jr., P.E.
Accident Research and Biomechanics, Inc.
27811 Avenue Hopkins, Suite 1
Valencia, CA 91354
arb@accidentresearch.com

Comparison of Pedestrian Subsystem Safety Tests Using Impactors and Full-Scale Dummy Tests

Yasuhiro Matsui, Adam Wittek and Atsuhiro Konosu
Japan Automobile Research Institute

ABSTRACT

Evaluation of car front aggressiveness in car-pedestrian accidents is typically done using sub-system tests. Three such tests have been proposed by EEVC/WG17: 1) the legform to bumper test, 2) the upper legform to bonnet leading edge test, and 3) the headform to bonnet top test. These tests were developed to evaluate performance of the car structure at car to pedestrian impact speed of 11.1 m/s (40 km/h), and each of them has its own impactor, impact conditions and injury criteria. However, it has not been determined yet to what extent the EEVC sub-system tests represent real-world pedestrian accidents.

Therefore, there are two objectives of this study. First, to clarify the differences between the injury-related responses of full-scale pedestrian dummy and results of sub-system tests obtained under impact conditions simulating car-to-pedestrian accidents. Second, to propose modifications of current sub-system test methods.

In the present study, the Polar (Honda R&D) dummy was used. This dummy was selected as it has been reported in the literature that it well represents motion of postmortem human subjects in lateral impacts. We impacted the dummy by one passenger car and one sport utility vehicle at a speed of 11.1 m/s. The results of the experiments using the Polar dummy were compared with those obtained using sub-system tests conducted according to EEVC/WG17 procedures. In this comparison, we analyzed the variables characterizing conditions of impact to the head (i.e., impact angle and impact speed of the head) and thigh/pelvis (bonnet edge deformation) as well as those describing the injury risk (i.e., knee shearing displacement, knee bending angle and tibia acceleration).

The present results suggest that the EEVC/WG17 headform and upper legform test procedures may overestimate severity of impact between car front and pedestrian head and pelvis, especially for sport utility vehicles. Furthermore, the current EEVC/WG17 legform impactor may not be suitable for evaluation of the aggressiveness of a high-bumper car front.

INTRODUCTION

In Japan, as a result of rapid motorization, pedestrian fatalities (persons dead within 24 hours) reached a peak of 5939 in 1970. After a decade, pedestrian fatalities reduced to almost half as shown in Figure 1 (ITARDA 2001). One of the countermeasures to reduce the fatalities was the establishment of an infrastructure whereby the sidewalk was separated from the roadway. Additionally, traffic safety education has been promoted. However, since 1980, annual pedestrian fatalities in Japan have remained between 2500 and 3000, and they still comprise 27% to 32% of all fatalities caused by traffic accidents. In other developed countries, pedestrian fatalities also remain high: approximately 7000 in the European Union (Davies 1997) and 5400 in the USA (IRTAD 1995).

For this reason, the European Enhanced Vehicle-safety Committee (EEVC) and International Standard Organization (ISO) have been considering test procedures to evaluate vehicle aggressiveness against pedestrians. Proposals for such procedures have focused mainly on subsystem tests because they seem less expensive, more simple and more repeatable than application of full-scale pedestrian dummies (Harris, 1989). Figure 2 summarizes the set-up of the subsystem test procedures proposed by the European Enhanced Vehicle-safety Committee/WG17 (EEVC/WG17, 1998). They were designed to evaluate performance of the car structure in an accident in which a passenger car traveling at a speed of 11.1 m/s hits a pedestrian crossing a roadway. In the EEVC/WG17 test procedures three different impactors representing head, thigh and leg are impacted against the car bonnet, bonnet-edge and bumper, respectively. However, it has not been determined yet to what extent these procedures represent real-world pedestrian accidents. For instance, in these procedures, one standard speed (11.1 m/s) of headform impactor is used, although several mathematical modeling and experimental studies have indicated that the head-to-bonnet impact speed varies depending on a vehicle shape and can be appreciably below the initial vehicle speed (EEVC/WG17, 1998).

Therefore, the present study focuses on the following two objectives. First, to clarify differences between the injury-related responses of a full-scale pedestrian dummy (Polar by Honda R&D) and results of sub-system tests obtained under impact conditions simulating car-to-pedestrian accidents. Second, to propose a modification of the current sub-system test method.

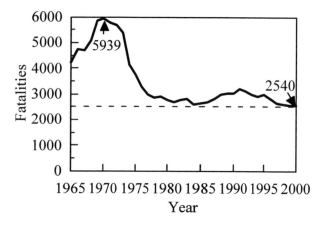

Figure 1. Pedestrian fatality trend in Japan. (ITARDA 2001)

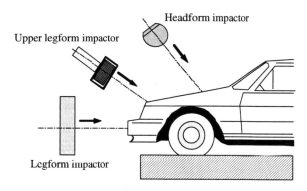

Figure 2. EEVC subsystem tests using three impactors.

METHODS

We conducted two types of tests: 1) EEVC legform and upper legform subsystem tests; and 2) full-scale dummy tests. In full-scale dummy tests, a car cut-body was placed on a hydraulically powered sled device and accelerated to a speed of 11.1 m/s. According to the EEVC/WG17, this speed corresponds to the impact severity of the subsystem tests.

Two different types of vehicles, compact and sport utility (SUV), were used here. Their specifications are summarized in Table 1. The compact vehicle had a front member beneath the bumper whereas the SUV did not.

The center of the bumper (D1) and center of the frontal bracket (D2) were selected as an impact location for the dummy leg and legform impactor tests as shown in Figure 3. The basis for selection of location D2 was that

it is likely to be the most aggressive area of a car front due to the large stiffness of the frontal bracket (Figure 4). For the upper legform test, the impact point was determined as the center of the contact area between the pelvis/thigh and bonnet leading edge observed in full-scale dummy tests.

Table 1. Specifications of tested vehicle.

Vehicle		Compact	SUV
Total length × total width × total height	mm	4180 × 1695 × 1375	3695 × 1695 × 1655
Minimum height	mm	150	205
Net weight	kg	960	1150
Gross weight	kg	1235	1370
Displacement	cc	1343	1998
Bumper material		resin	resin

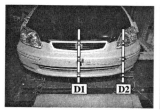 

Compact car   SUV

Figure 3. Impact locations used in the present study.

Compact car   SUV

Figure 4. Structure beneath bumper at impact location D2.

SUBSYSTEM TEST

Legform to Bumper Test – Figure 5 shows the legform – to–bumper test conducted at the Japan Automobile Research Institute (JARI). The vehicle cut-body was secured to the ground to eliminate the effects of frontal suspension. A legform impactor developed by the Transport Research Laboratory (TRL) (Lawrence et al. 2000) was used as the test device. By means of a set of sensors commercially installed on this impactor we measured knee shearing displacement, knee bending angle and tibia acceleration. The data were sampled at 10 kHz and processed by means of an SAE 180 filter. The motion of the legform impactor was recorded by high-speed digital camera (500 frames/second).

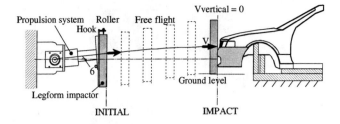

Figure 5. Subsystem test setup of legform-to-bumper test conducted at JARI.

Upper Legform–to–Bonnet Leading Edge Test – Figure 6 shows the upper legform impact test conducted at JARI. The vehicle body was fixed to the ground. The upper legform impactor developed by TRL (Lawrence et al. 1991 and Hardy 1993) was used. We measured the impact force and bending moment, which is a typical procedure in tests using an upper legform impactor. The data were sampled at 10 kHz and processed by means of the SAE 180 filter.

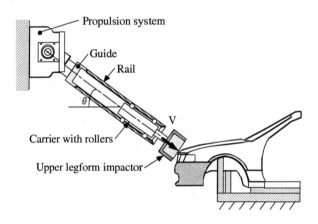

Figure 6. Subsystem test setup of upper legform-to-bonnet leading edge test conducted at JARI.

Impact conditions, such as speed, impact angle and impact energy were determined from the look-up graphs proposed by EEVC/WG17 (1998) (Table 2). The effective impactor mass M (kg) was calculated using the following equation:

$$M = \frac{2E}{V^2}, \qquad (1)$$

where V (m/s) is the impact speed and E (J) is the impact energy.

It should be noted here that the EEVC/WG17 test procedure does not require upper legform impact tests when the energy E is lower than 200 J. However, for the compact car, we performed such tests at the energy of 82 and 121 J to collect the data necessary to compare this procedure with the results of full-scale dummy impact.

Table 2. Upper legform impact conditions determined using EEVC/WG17 (1998) look-up graphs.

Vehicle		Compact		SUV	
Location		Center	Side	Center	Side
Bonnet leading edge height	mm	621	646	880	885
Bumper lead	mm	140	140	151	166
Impact energy	J	82	121	700	700
Impact speed	m/s	4.2	5.1	11.11	11.11
Impactor mass	kg	9.155	9.155	11.340	11.340
Impact angle	degree	43	43	34	34

FULL-SCALE DUMMY TEST

To understand the human dynamic and injury-related responses in a car-pedestrian accident, we conducted impact tests using the full-scale dummy (Polar) developed by Honda R&D (Akiyama et al. 2000). To the best of our knowledge, it is the most advanced and biofidelic pedestrian dummy currently available.

The height of the Polar dummy is 1775 mm and the weight is 770 N. Its femoral condyles have a human-like ellipsoidal shape, and the knee ligaments are represented by four cables connected with a system of springs and rubber tubes. The dummy was suspended from the roof in the walking posture as shown in Figure 7, and released 100 ms prior to the impact to ensure that its whole weight loads the lower extremities at the time when a car cut-body hits the dummy. The position of the impacted left leg was adjusted to meet the impact locations (D1 and D2) described in Figure 3. However, the location at which contact occurred between the dummy lower extremity and the bonnet leading edge differed between the compact car and the SUV. In the tests using the compact car, the leading edge impacted the thigh, whereas the SUV impacted the pelvis (Figure 8).

Figure 7. Setup of full-scale test conducted at JARI.

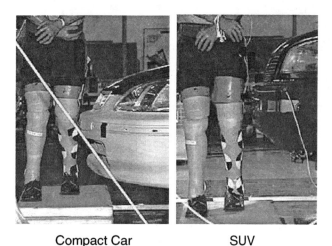

Compact Car   SUV

Figure 8. Contact location of full-scale dummy depends on vehicle type.

<u>Measured Variable</u> – To compare acceleration of the dummy leg with that obtained in subsystem tests using the TRL legform impactor, an additional accelerometer was equipped 66 mm below the knee joint center on the left (i.e., impacted) leg. The signal collected by this accelerometer was sampled at 10 kHz and processed by means of the SAE 180 filter.

Key dimensions of the Polar dummy lower extremity and sensor position are summarized in Figure 9.

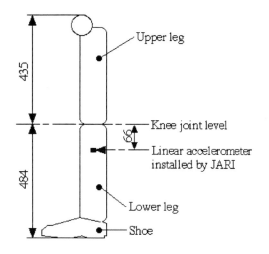

Figure 9. Dimensions of full-scale dummy left leg.

<u>Dummy Motion Analysis</u> – Motion of the dummy was recorded by means of a high-speed digital camera (500 frames/second). To determine dummy head impact speed, head impact angle, thigh speed, knee shearing displacement and knee bending angle, photographic targets were placed on the dummy head, torso and lower extremities (Figure 10). Motion of the targets was traced and analyzed using Image Express workstation (NAC, 1995).

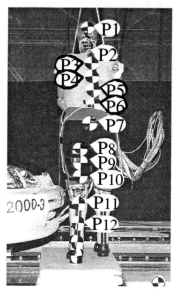

Figure 10. Photographic targets placed on Polar dummy. The head target P1 was placed near the head gravity center CG.

<u>Wrap Around Distance, Head Impact Speed and Head Impact Angle</u> – The wrap around distance (WAD) was defined as the distance from the ground to the center of the bonnet area that deformed due to contact with the dummy head (Figure 11a).

The head speed V_{HR} was determined as the magnitude of the relative velocity between target P1 and a car. When determining the head speed, it was assumed that the car cut-body did not move in the vertical direction because it was firmly secured to the sled:

$$V_{HR} = \sqrt{V_{Hy}^2 + V_{Hz}^2},$$
$$V_{Hy} = V_{P1y} - V_{car}, \; V_{Hz} = V_{P1z}. \tag{2}$$

Head velocity angle δ_H was defined as the angle between the velocity vector of target P1 and the horizontal direction (Figure 11b). The basis for this definition was that the EEVC headform test procedure determines the impactor orientation in relation to the horizontal plane. The head impact speed and head impact angle were defined as values of the head speed and head velocity angle at the time of initial contact between the dummy head and bonnet. The time of beginning of head-bonnet contact was determined from high-speed digital camera recordings. As the recording was done with a frequency of 500 frames/second, the resolution of this determination was not higher than 2 ms.

a)

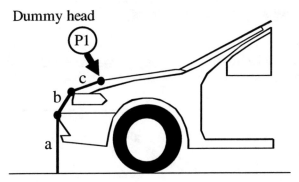

Wrap Around Distance (WAD) = a+b+c

b)

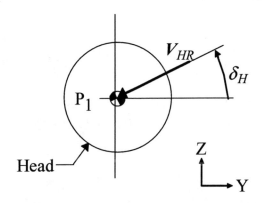

Figure 11. a) Definition of wrap around distance (WAD), and b) Head velocity vector V_{HR} and head velocity angle δ_H.

Thigh/Pelvis Speed – The bonnet leading edge of the compact car impacted the dummy thigh, whereas the bonnet leading edge of the SUV impacted the pelvis. Therefore, for the analysis, we selected the thigh (target P9) and pelvis (target P7) speeds obtained in the experiments using the compact car and SUV, respectively. The thigh/pelvis speed V_I is a magnitude of vector component of relative velocity V_R between the target P9/P7 and the car in the direction perpendicular to the thigh/pelvis longitudinal axis (Figure 12). As with the head speed, when calculating the thigh/pelvis speeds, we assumed that the car cut-body did not move in the vertical direction. These speeds were determined using the following formulae:

$$V_I = |V_R \cos \gamma|,$$
$$\gamma = \beta - \delta,$$

(3)

where β is the thigh/pelvis velocity angle (i. e. , angle between the thigh/pelvis velocity V_I and horizontal direction), and δ is the angle between horizontal direction and velocity of target P9/P7 in relation to a car.

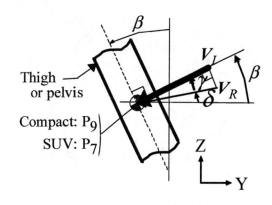

Figure 12. Definition of thigh/pelvis velocity vector V_I and velocity angle β.

Knee Shearing Displacement (D_s) and Knee Bending Angle – The dummy knee shearing displacement and knee bending angle were expressed in relation to their initial values, which were set to zero at the start of dummy-car impact. They were calculated based on the relative movement of targets P9-P10 and P11-P12 (Figure 13):

$$D_S = Y_{P11} + d_L \sin \alpha - Y_{P10} - d_T \sin \beta,$$
$$\alpha = \arctan \left(\frac{Y_{P11} - Y_{P12}}{Z_{P11} - Z_{P12}} \right),$$
$$\beta = \arctan \left(\frac{Y_{P10} - Y_{P9}}{Z_{P9} - Z_{P10}} \right),$$

(4)

Knee bending angle $= \alpha + \beta$.

Thus, the set of equations in Eq. (4) determines shearing displacement of the dummy knee joint as the horizontal displacement between its femoral condyles and the tibial plateau in the lateral plane. As the dummy lower extremities exhibited complex three-dimensional motion during an impact, this approach may yield results that do not exactly correspond to those obtained by means of a specialized potentiometer installed in the TRL legform impactor. Therefore, the present analysis is limited to comparison of general tendencies in D_s-time histories of Polar dummy and shearing displacement of the TRL legform impactor.

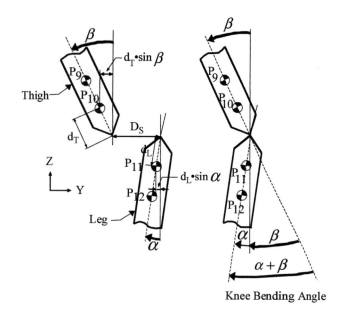

Knee Bending Angle

Figure 13. Determination of knee shearing displacement and knee bending angle using motion of targets P9, P10, P11 and P12.

COMPARISON OF UPPER LEGFORM IMPACTOR TESTS AND FULL-SCALE DUMMY TESTS

To understand the differences in severity of the full-scale dummy tests and those conducted according to the EEVC/WG17 upper legform impactor test procedure, we compared the bonnet edge residual deformations obtained in each of these tests. The basis for this approach is that such deformation is a function of impact energy.

RESULTS

KINEMATICS OF PEDESTRIAN HEAD

<u>Head Trajectory</u> – Both horizontal and vertical displacements of the photographic target located on the dummy head were larger in the experiments using the compact car than the SUV (Figure 14). The main reason for these differences in the head target trajectories (referred to as head trajectories) between these two cars is that the SUV bonnet height is greater than that of compact car (Table 2). This height implies a shorter distance between the head gravity center CG and the instantaneous axis of dummy rotation for the SUV than for the compact car as the dummy rotated about its contact area with the bonnet leading edge. Furthermore, the high bonnet leading edge caused the head-to-bonnet contact to occur earlier for the SUV than the compact car as will be explained in the Head Speed section. This earlier contact resulted in shorter vertical displacement of the head CG in the experiments using the SUV.

In addition to the just discussed differences related to car type (i.e., car front shape), the current results suggest that the head trajectories can depend also on the impact location. For a compact car, the head horizontal displacement was clearly larger in the experiments using impact location D2 (frontal bracket) than those in which the bonnet center was impacted. One possible explanation for this result can be the "rounded" shape of this car front.

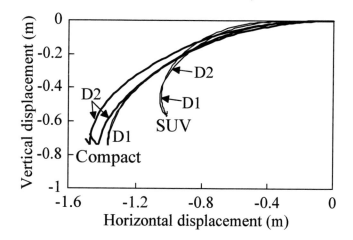

Figure 14. Head trajectory. Both horizontal and vertical displacements of the head target P1 were calculated in relation to a car.

<u>Wrap Around Distance (WAD)</u> – Table 3 summarizes WAD and the ratio of the WAD and pedestrian dummy height (1775 mm). For the compact car, this ratio ranged from 1.05 to 1.09, and was clearly larger than the 0.96 observed in the SUV experiments. This result is consistent with the differences in the head trajectories between the SUV and the compact car discussed in the previous section.

Table 3. WAD and ratio between WAD and pedestrian dummy height.

Vehicle	Compact		SUV	
Impact location	D1	D2	D1	D2
Number of impact tests	2	2	1	1
WAD (mm)	1850 1870	1940 1930	1700	1700
Average WAD (mm)	1860	1935	1700	1700
Average WAD / Dummy height*	1.05	1.09	0.96	0.96

* Dummy height: 1775 mm.

44

Head Speed – As with the head trajectories, the present results indicate appreciable differences in the head speed between the SUV and the compact car. For the SUV, the maximum head speed was clearly lower than that observed when the dummy was impacted by the compact car. Simultaneously, the contact between the head and bonnet occurred around 20 ms earlier with the SUV than with the compact car (Figure 15). Consequently, the average head impact speed (i.e., head speed at the start of contact between the head and bonnet) was 8.0 m/s (around 29 km/h) for the SUV, and 8.9 m/s (around 32 km/h) for the compact car (Table 4). Thus, the average ratio of the head impact speed to the initial car speed (around 11.1 m/s), referred to as the average head impact speed ratio, was lower for the SUV (around 0.7) than for the compact car (around 0.8) (Table 4), for much the same reason given when presenting the head trajectories. The high position of the SUV bonnet leading edge implies a shorter distance between the instantaneous axis of dummy rotation and the head CG, which resulted in the lower head impact speed for SUV than the compact car.

It should be also noted that in all our experiments the head impact speed was lower than the initial car speed (i.e., 11.1 m/s), i.e., the head impact speed ratio was less than 1.0 (Table 4). On the other hand, in the EEVC/WG17 headform impactor test procedure, one standard impact speed of 11.1 m/s is used for all the car shapes and categories. Thus, our results suggest that this test procedure may overestimate the severity of the impact between the bonnet and the pedestrian head, especially in accidents involving SUV. The headform impactor speed in this test procedure should be in 20%-30% lower than the initial car speed.

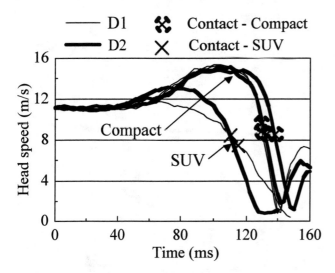

Figure 15. Head speed. This speed was calculated in relation to a car.

Table 4. Head impact speed and the ratio between head impact speed and initial car speed.

Vehicle	Compact		SUV	
Impact location	D1	D2	D1	D2
Number of impact tests	2	2	1	1
Head impact speed (m/s)	8.8	8.5	7.5	8.4
	9.7	8.4		
Average head impact speed (m/s)	8.9		8.0	
Average head impact speed ratio*	0.80		0.72	

*Average head impact speed / initial car speed

Head Impact Angle – In the experiments in which the dummy was impacted by the SUV, the average head impact angle was around 94° which is clearly larger than the 80° angle observed in impacts using the compact car (Table 5). Thus, in our experiments, the head velocity vector at the start of contact between the SUV bonnet and dummy head was almost vertically oriented. On the other hand, the EEVC/WG17 test procedure using an adult headform impactor assumes an angle of 65° from the horizontal plane for all the car types.

Table 5. Head impact angle at the time of contact with bonnet or windshield.

Vehicle	Compact		SUV	
Impact location	D1	D2	D1	D2
Number of impact tests	2	2	1	1
Head impact angle (degree)	81	82	88	100
	80	75		
Average head impact angle (degree)	80		94	

LOWER EXTREMITY

Dynamic Sequence And Motion Pattern

Compact Car – In the experiments conducted using the compact car, the legform impactor leg rebounded from the bumper starting from around 20 ms after the beginning of contact between the leg and bumper. On the other hand, the dummy leg remained in contact with the bumper (Figure 16). Furthermore, in the subsystem tests using upper legform impactor, the impact angle is virtually constant as the impactor motion is constrained by the guide rails (Figure 6). In contrast, the dummy thigh rotated, and its impact direction varied with time.

SUV – In the experiments using the SUV, the bumper impacted the dummy thigh and the bonnet leading edge impacted the pelvis (Figure 17). This is in contrast to the experiments conducted on the compact car, in which the leg and thigh were hit. Around 30 ms after the initial contact with the bumper, the legs of both the legform impactor and dummy tended to intrude under the bumper. Simultaneously, the legform impactor thigh tended to rebound from the bumper, whereas the dummy pelvis leaned on the bonnet leading edge.

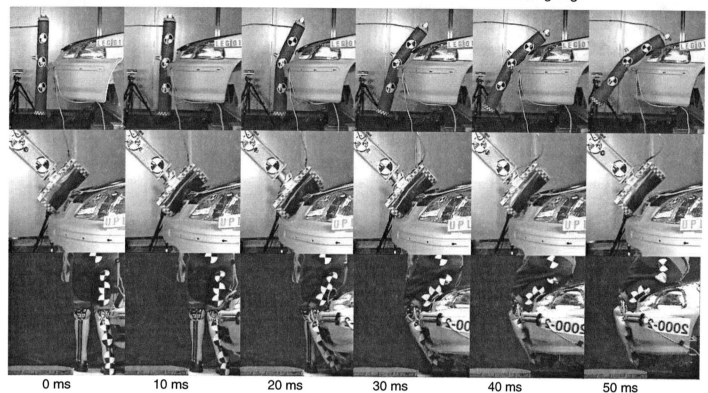

| 0 ms | 10 ms | 20 ms | 30 ms | 40 ms | 50 ms |

Figure 16. Comparison of dynamic motion of the subsystem impactors (legform – first row, and upper legform – second row) and full-scale dummy lower extremity (third row) during impact with center (impact location D1) of compact car.

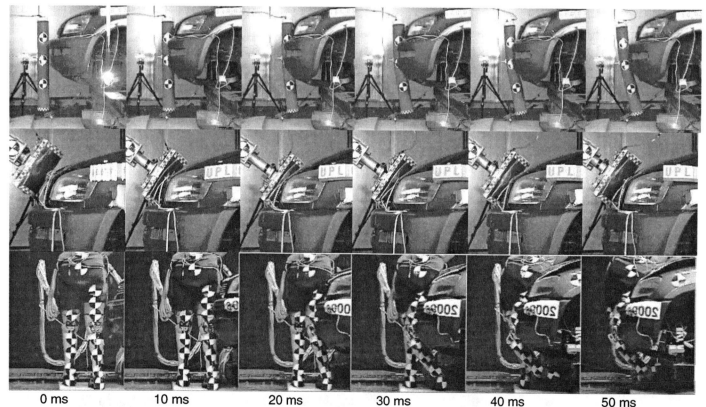

| 0 ms | 10 ms | 20 ms | 30 ms | 40 ms | 50 ms |

Figure 17. Comparison of dynamic motion of the subsystem impactors (legform – first row, and upper legform – second row) and full-scale dummy lower extremity (third row) during impact with center (impact location D1) of SUV.

Thus, the present comparison of the motion pattern of the TRL legform impactor and Polar dummy subjected to lateral impacts representing car-pedestrian accidents suggests the following. In the initial impact phase, the key features of the kinematic responses of the impactor, such as magnitude of the knee joint angle, seemed to be similar to that of the dummy, especially for compact car. Starting from around 20-30 ms of impact, differences in legform impactor and dummy responses could be observed. They were more evident for the SUV than for the compact car. This might be related to the effects of inertia of the dummy torso which pulled the thigh and pelvis towards the bonnet.

Knee Shearing Displacement – Time histories of the knee shearing displacement are presented only for the initial 20 ms of impact since injuries related to lateral displacement between femoral condyles and tibia plateau (i.e., shearing-type injuries) typically occur in the initial impact phase (Kajzer et al., 1997, 1999).

The present study suggests differences between results of the legform impactor and full-scale dummy tests. When the legform impactor impacted the SUV, its shearing displacement was lower than that obtained in impacts using the compact car (Figures 18 and 19). However, no such tendency was observed for the Polar dummy (Figures 20 and 21).

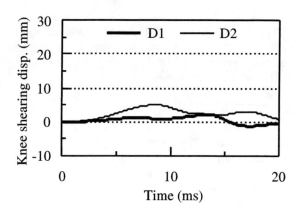

Figure 19. Knee shearing displacement of legform impactor upon contact with SUV.

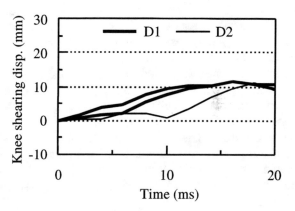

Figure 20. Knee shearing displacement of full-scale dummy upon contact with compact car.

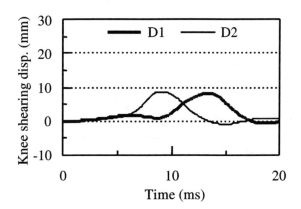

Figure 18. Knee shearing displacement of legform impactor upon contact with compact car.

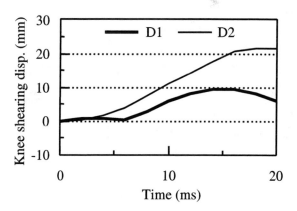

Figure 21. Knee shearing displacement of full-scale dummy upon contact with SUV.

47

Knee Bending Angle – During the initial 15 ms of impact with the compact car, the knee bending-angle time histories obtained using Polar dummy were close to those observed for the TRL legform impactor (Figures 22 to 23). However, between 20 and 30 ms, important differences in "bending-type" responses of these two test devices were found. Within this time-window, the knee bending angle of the TRL legform impactor either saturated or reached its maximum, whereas that of the dummy continuously increased.

In impacts against the SUV, the behavior of knee bending-angle time histories obtained using Polar dummy were different from those observed for TRL legform impactor (Figures 24 to 25). At location D1, the maximum knee bending angle of the TRL legform impactor was around 9°. Simultaneously, the knee bending angle of the pedestrian dummy exceeded 30° (Figure 24). The likely reason for this phenomenon is disregarding the effects of torso and head inertia on the lower extremity responses in the EEVC/WG17 legform impactor. Therefore, it cannot be ruled out that the TRL legform impactor might underestimate the risk of injury to knee joint ligaments in impacts against high-bumper cars. This, in turn, suggests that this impactor may not be suitable for estimation of aggressiveness of a high-bumper car front. Therefore, for the estimation of SUV (i.e., high-bumper car) frontal bumper aggressiveness, equipping a legform impactor with equivalent mass simulating the effects of inertia of pedestrian upper body should be considered.

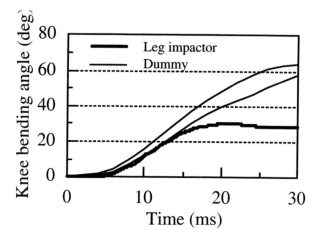

Figure 23. Knee bending angle obtained from full-scale dummy and legform impactor upon contact with side (impact location D2) of compact car.

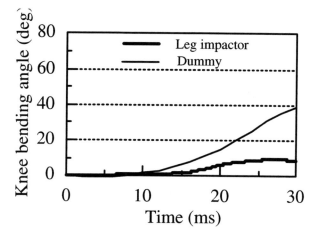

Figure 24. Knee bending angle obtained from full-scale dummy and legform impactor upon contact with center (impact location D1) of SUV.

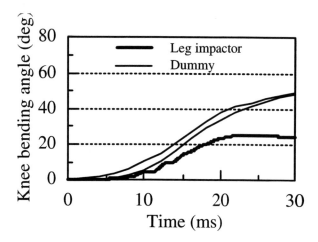

Figure 22. Knee bending angle obtained from full-scale dummy and legform impactor upon contact with center (impact location D1) of compact car.

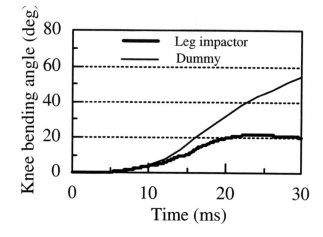

Figure 25. Knee bending angle obtained from full-scale dummy and legform impactor upon contact with side (impact location D2) of SUV.

Leg Acceleration – The general behavior of the time histories of the leg acceleration obtained from the full-scale dummy and legform impactor were similar for both the compact car and the SUV (Figures 26 to 29). However, the peak acceleration obtained from the legform impactor was approximately double in all our experiments. When interpreting this result one should realize that although the Polar dummy is only a full-scale mechanical model of the 50-percentile pedestrian, its biofidelity in terms of lower extremity responses is likely to be higher than that of the TRL legform impactor. The reason for this suggestion is that the impactor leg is virtually rigid, whereas that of the Polar dummy can simulate deformation of the human tibia. It has been shown in mathematical modeling study by Konosu et al. (2001a) that human leg acceleration, in a given measurement point, cannot be appropriately represented by means of legform impactor with rigid leg.

However, the present study alone is insufficient to verify the validity of the hypothesis that the rigid leg in the TRL legform impactor is the only reason for the present differences in the leg acceleration-time histories of this impactor and Polar dummy. One cannot completely rule out the possibility that some of these differences could be related to different properties of padding of the dummy leg (Confor™ foam covered by the PVC "skin") and legform impactor (Confor™ foam covered by the neoprene "skin").

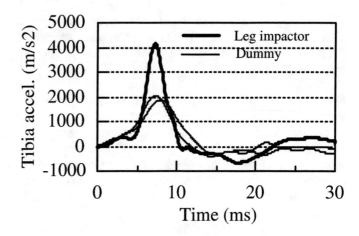

Figure 27. Leg acceleration obtained from full-scale dummy and legform impactor upon contact with side (impact location D2) of compact car.

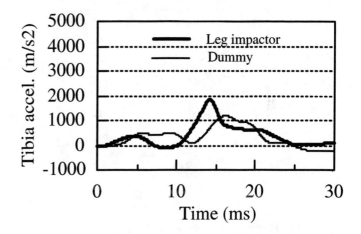

Figure 28. Leg acceleration obtained from full-scale dummy and legform impactor upon contact with center (impact location D1) of SUV.

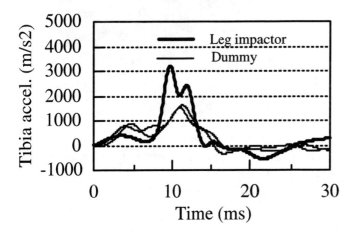

Figure 26. Leg acceleration obtained from full-scale dummy and legform impactor upon contact with center (impact location D1) of compact car.

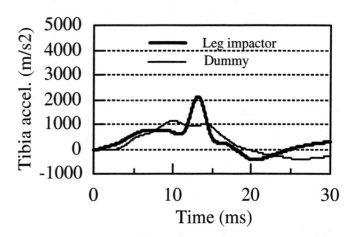

Figure 29. Leg acceleration obtained from full-scale dummy and legform impactor upon contact with side (impact location D2) of SUV.

Thigh Speed – Speed of the dummy thigh appreciably decreased during impact. The rate of this decrease might differ according to the impact location, i.e., D1 and D2 (Figures 30 to 31).

Magnitudes of the thigh/pelvis speed-time histories either yielded zero in the relatively late impact phase or saturated at the level slightly exceeding zero. One possible explanation for this phenomenon might be that these speeds were calculated as relative speeds of photographic targets located on the thigh/pelvis longitudinal axis. These targets do not represent instantaneous contact points between the bonnet leading edge and dummy thigh/pelvis.

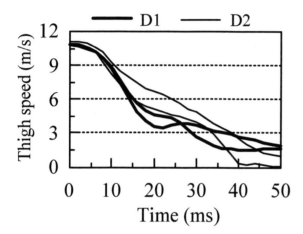

Figure 30. Thigh speed obtained from full-scale dummy upon contact with center (D1) and side (D2) of compact car.

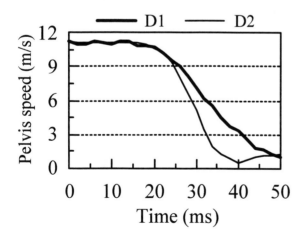

Figure 31. Pelvis speed obtained from full-scale dummy upon contact with center (D1) and side (D2) of SUV.

COMPARISON OF UPPER LEGFORM IMPACTOR TESTS AND FULL-SCALE DUMMY TESTS

The residual deformation of the bonnet edge of the compact car was similar in the upper legform impactor and full-scale dummy tests since energy of the upper legform test was low (below 200 J) (Figures 32 and 33).

Figure 32. Residual deformation on bonnet leading edge at side (impact location D2) of compact car after contact with full-scale dummy.

Figure 33. Residual deformation on bonnet leading edge at side (impact location D2) of compact car after impact by upper legform impactor.

On the other hand, for the SUV, energy of the upper legform impactor test determined from the EEVC look-up graphs was 700 J which is a rather large value. In consequence, with the SUV, the residual deformation obtained in the upper legform impactor test clearly exceeded that observed when the side of SUV front impacted Polar dummy (Figures 34 and 35). These results suggest that impact energy of the upper legform impactor test seems to be appropriate when evaluating safety performance of a compact vehicle front. However, it might be too large for a SUV.

Figure 34. Residual deformation on bonnet leading edge at side (impact location D2) of SUV after contact with full-scale dummy.

Figure 35. Residual deformation on bonnet leading edge at side (impact location D2) of SUV after impact by upper legform impactor.

DISCUSSION

Generally, the method of subsystem testing has been based on the car shape and accident data dating back over twenty years ago. Pedestrian injury tolerance does not change. However, the contact conditions between pedestrian and car have changed, because the recent frontal shape of the car, especially for sport utility vehicles, has undergone a significant change compared to the previously designed target. The actual contact condition should be used when establishing the pedestrian test, especially for sport utility vehicles.

In this research, only two types of vehicles (mid-nineties models) were used. Thus, one should be careful when generalizing our findings. For instance, in many currently designed vehicles, the bonnet length has become short. In such vehicles the head is likely to contact the windshield or the front roof part. The contact would occur earlier than for the vehicles we used. Thus, both the head impact speed and angle would change in comparison to those obtained in the present study.

Therefore, to establish reliable test methods for pedestrian protection, we need to understand the conditions of impact between the pedestrian head and vehicles of various shapes and types.

Impact energy of the subsystem for the upper legform impactor was set with the 400 J reduction (700 J) from the calculated value (1100 J) in the look-up graph, because of the application of the upper limit energy defined by the EEVC/WG17. However, the residual deformation of the bonnet leading edge of the sport utility vehicle caused by the contact with the upper legform impactor with 700 J was still greater than that by a full-scale dummy (Figures 34 and 35). The likely reason for this is that the upper legform impactor is constrained from being propelled into the bonnet leading edge; consequently, the impact direction does not change, and more energy is applied to the area. Furthermore, the impact energy for the EEVC/WG17 upper legform impactor procedure was determined from the forces and relative displacement of thigh into the bonnet obtained from full-scale tests (Lawrence, 1998). However, the present study clearly shows that when a 50-percentile pedestrian is impacted by a sport utility vehicle, the contact indeed occurs between pelvis and bonnet leading edge.

Thus, the present study suggests that in the EEVC/WG17 look-up graph for the upper legform impactor test, the impact energy for a car with a high bonnet leading edge, such as in the case of a sport utility vehicle, might be overestimated. In this vehicle type, decreasing the energy should be considered to meet the real-world pedestrian impact condition.

The behavior of the full-scale dummy lower–extremity varied due to the car shape (Figures 36 to 39). Fundamentally, the test procedure with the legform impactor was established to assess the risk of knee joint ligament injury and tibia fracture. Due to the effects of car shape on the lower extremity responses, it might not be appropriate to use the legform impactor to evaluate aggressiveness of a vehicle with high bumper height, a suggestion similar to the one made in the EEVC/WG17 (1998) report. The height limitation should be determined based on the proper method including the use of computer simulation.

Furthermore, our results indicated important differences between the knee bending angle-time histories of the TRL legform impactor and the Polar dummy. In impacts against SUV, the knee bending angle of the impactor was appreciably lower than that of the dummy (Figures 24 and 25). However, in impacts against compact car the differences between the knee bending angle-time histories of the Polar dummy and TRL legform impactor seemed to be relatively low during the initial 15 ms, and then increased as the impactor angle saturated (Figures 22 and 23). In consequence, in impacts against a compact car, the magnitude of the knee bending angle-time histories of both the TRL legform impactor and Polar dummy exceeded the value of 15° proposed by the EEVC (1998) as the injury threshold for knee joint ligaments. However, it may not be correct to interpret this result as an indication that the dummy and impactor predicted a similar risk of ligament injury. The reason for this suggestion is that the basis for the EEVC (1998) threshold for knee ligament injury is somewhat uncertain as it was derived based on the assumption that the legform impactor exactly represents the human lower extremity responses. Unfortunately, this assumption does not apply to the current TRL legform impactor as the leg of this impactor is virtually rigid, whereas the human one appreciably deforms when impacted by a car. It has been recently shown in the mathematical modeling study by Konosu et al. (2001b) that to compensate the effects of infinite stiffness of the impactor leg on magnitude of the impactor knee bending angle, the injury threshold should be increased from 15° to around 30°. Our experimental results clearly indicate that the current TRL legform impactor is not capable of evaluating the injury risk to knee joint ligaments when the threshold value is close to 30°. Thus, modifications of the TRL legform impactor may be needed in order to take into account the recent findings on the effects exerted by the infinite stiffness of this impactor leg on the evaluation of injury risk to knee ligaments.

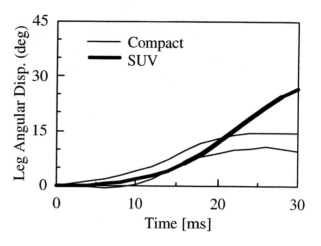

Figure 37. Leg angular displacement α obtained from full-scale dummy tests by central impact (impact location D1).

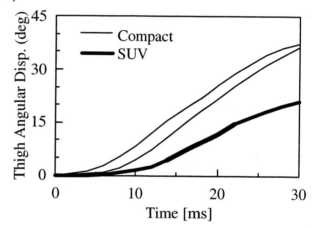

Figure 38. Thigh angular displacement β obtained from full-scale dummy tests by side impact (impact location D2).

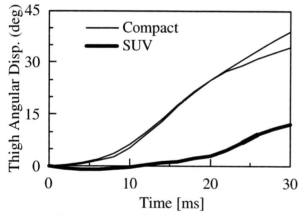

Figure 36. Thigh angular displacement β obtained from full-scale dummy tests by central impact (impact location D1).

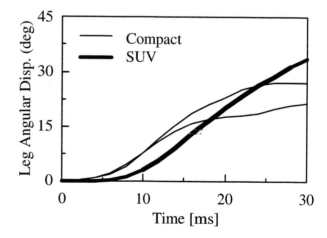

Figure 39. Leg angular displacement α obtained from full-scale dummy tests by side impact (impact location D2).

Several areas for improvement of the present analysis should thus include:

1. Computer simulation using an FEM full-scale model should be applied to analyze in detail and validate the impact conditions of the subsystem upper legform impact test procedure.

2. A mid-sized passenger car should be used to compare of the EEVC/WG17 upper legform impactor test procedure and full-scale dummy impact situation. The reason for this suggestion is that in the present study, we used only a compact car (impact energy below 200 J) and sport utility vehicle (impact energy over 700 J).

3. The type of pedestrian injury sustained when he/she is hit by a sport utility vehicle should be identified based on the accident investigation. We should also consider the proper test method for evaluation of the injury risk caused by the bumper and bonnet leading edge of a sport utility vehicle.

4. The impact condition of the chest should be obtained from the dynamic sequence of the current full-scale dummy result, because in pedestrian accidents in Japan, the most common fatal injuries involve the head (61%) and chest (11%) (Matsui 2001).

CONCLUSIONS AND RECOMMENDATIONS FOR MODIFICATIONS OF PEDESTRIAN SUB-SYSTEM TEST PROCEDURES

Sub-system tests and full-scale dummy tests using compact car and sport utility vehicle (SUV) were conducted to determine their differences and to propose a modification of sub-system tests for recent passenger cars. The following conclusions were made based on these tests.

1. The ratio of head impact speed to car initial speed was lower for SUV (around 0.7) than for compact car (around 0.8). On the other hand, in the EEVC/WG17 headform impactor test procedure, one standard impact speed of 11.1 m/s is used. Thus, this test procedure may overestimate the severity of impact between the bonnet and the pedestrian head, especially in accidents involving SUVs.

2. The head impact angle was larger for SUV than for the compact car. On the other hand, the EEVC/WG17 test procedure using an adult headform impactor assumes the same angle for all car types.

3. The current TRL legform impactor may not be suitable for evaluation of the aggressiveness of a high-bumper car front.

4. For the SUV, the residual deformation of the bonnet leading edge after the impact using the upper legform impactor was larger than for a full-scale dummy. Thus, the subsystem test method using the upper legform impactor for the estimation of SUV frontal aggressiveness should be reconsidered.

Based on these conclusions we suggest the following modifications of the pedestrian subsystem test procedures:

1. When evaluating frontal aggressiveness of compact car and SUV in terms of risk of pedestrian head injury, the headform impactor speed should be in 20%-30% lower than the initial car speed. Furthermore, lower impact speed and larger impact angle should be used for SUV than compact car.

2. For the estimation of SUV frontal bumper aggressiveness, equipping a legform impactor with equivalent mass simulating the effects of inertia of pedestrian upper body should be considered.

3. For the estimation of SUV frontal bonnet leading edge aggressiveness, decreasing the upper legform impactor energy should be considered.

ACKNOWLEDGMENTS

The current research was financially supported by the Japan Automobile Manufacturers Association.

The authors thank Honda Inc. for generously providing the Polar dummy and assisting in the full-scale dummy experiments.

REFERENCES

1. Akiyama A., Okamoto M., and Rangarajan N. (2001) Development and Application of the New Pedestrian Dummy. Proc. 17th International Conference on the Enhanced Safety of Vehicles, Amsterdam, The Netherlands, Paper No. 463.

2. Davies R. G., Clemo L. C. and Bacon D. G. C. (1997) Study of Research into Pedestrian Protection MIRA, European Commission.

3. European Enhanced Vehicle-safety Committee (1998) Improved Test Methods to Evaluate Pedestrian Protection Afforded by Passenger Cars. EEVC Working Group 17 Draft Report.

4. Hardy B. (1993) Notes on the Use of the TRL Prototype Pedestrian Bonnet Leading Edge Impactor, Version 1.1 TRL document.

5. Harris, J. (1989) A Study of Test Methods to Evaluate Pedestrian Protection for Cars. Proc. 12th International Conference on the Enhanced Safety of Vehicles, pp. 1217-1225.

6. Institute for Traffic Accident Research and Data Analysis of Japan (2001) Annual Traffic Accident Report in 2000 (in Japanese). Tokyo.

7. International Road Traffic and Accident Database (1995).

8. Kajzer J., Schroeder G., Ishikawa H., Matsui Y. and Bosch U. (1997) Shearing and Bending Effect at the Knee Joint at High Speed Lateral Loading. SAE-Paper No. 973326, Proc. 41st STAPP Car Crash Conference, pp. 151-165.

9. Kajzer J., Matsui Y., Ishikawa H., Schroeder G. and Bosch U. (1999) Shearing and Bending Effect at the Knee Joint at Low Speed Lateral Loading. SAE-Paper No. 1999-01-0712, SAE Transactions Section 6, pp. 1159-1170.

10. Lawrence G., Hardy B. and Harris J. (1991) Bonnet Leading Edge Subsystem Test for Cars to Assess Protection for Pedestrians, Proc. 13th Enhanced Safety Vehicle.

11. Lawrence G., (1998) The Upper Legform Test Explained and Justified, Transport Research Laboratory, Paper of EEVC/WG17 Committee.

12. Matsui Y. (2001) Biofidelity of TRL Legform Impactor and Injury Tolerance of Human Leg in Lateral Impact, Proc. 45th STAPP Car Crash Conference.

13. Konosu, A., Ishikawa, H., Tanahashi, M. (2001a) Reconsideration of Injury Criteria for Pedestrian Subsystem Legform Test – Problems of Rigid Legform Impactor, Proc. 17th International Conference on the Enhanced Safety of Vehicles, Amsterdam, The Netherlands, Paper No. 263.

14. Konosu, A., Ishikawa, H., Tanahashi, M. (2001b) Reconsideration of Injury Criteria for Pedestrian Subsystem Legform Test (in Japanese). JARI Research Journal Vol.23 No.9.

15. NAC (1995) Image Express Users Manual for Application Version 1.26, NAC Incorporated, Woodland Hills, California, USA.

16. Transport Research Laboratory (2000) TRL Pedestrian Legform Impactor User Manual. Version 2.0.

Head Injuries in Vehicle-Pedestrian Impact

Koji Mizuno
Traffic Safety and Nuisance Research Institute

Janusz Kajzer
Nagoya University

ABSTRACT

In vehicle-pedestrian impacts, the kinematics and severity of pedestrian injuries are affected by vehicle front shapes. Accident analyses and multibody simulations showed that for mini vans the injury risk to the head is higher, while that to the legs is lower than for bonnet-type cars. In mini-van pedestrian impacts, pedestrians ran high risks of a head impact against stiff structures such as windshield frames.

When pedestrians are struck by a car with a short hood length, their heads are likely to strike into or around the windshield. The injury risks to the head by such an impact were examined by headform impact tests. The HIC rises from contact with the cowl, windshield frame or A pillar, and it lessens with increasing distance from these structural elements.

INTRODUCTION

In a vehicle-pedestrian impact, the pedestrian injury risks depend on the vehicle shape since together with the impact locations and body velocities the vehicle shape determines the severity of resulting injuries. Some publications have discussed the effects of the shape of cars and their bumper height, bumper lead and hood edge height on the injury risk to pedestrians [1][2][3]. Although, the number of mini vans and the SUVs (Sports Utility Vehicle) are increasing, pedestrian behavior and the injury risk to each body region have not been assessed for these types of vehicles.

To evaluate the injury risk to each body region, sub-system tests using impactors are carried out. The European Experimental Vehicle Committee (EEVC) proposed three sub-system tests: headform impactor to bonnet top, legform to bumper, and upper legform to bonnet leading edge [4]. The International Standard Organization also presented a similar sub-system tests.

Head injuries cause a serious threat to life and recovery is often incomplete. Therefore, it is most important to evaluate injury risks to the head. The EEVC test method prescribes that the head impact test shall be made with the bonnet top within the boundaries defined as a wrap around distance (WAD) of 1500 mm and 2100 mm at a velocity of 40 km/h [4]. In this test method, the windscreen and A pillars are excluded from the test area. The EEVC presented these test methods in its first report of EEVC WG10 [5]. However, when these pedestrian test methods were firstly discussed, most cars had an upright frontal area and a long hood. Since modern cars have become smaller with a short and steep bonnet, the head impact locations have changed from the hood to the cowl or windscreen in actual accidents. Thus, it was suggested that the injury risks to the head by contact on and around the windscreen should be investigated [4][6].

In this study, injury risks to the pedestrians upon impact against various shapes of vehicles will be investigated. Using accident analysis and mathematical simulations, such injury risks are compared for two different vehicle shapes, i.e., the sedan and the mini van. Since pedestrian head impact locations vary depending on the vehicle's front shape and the pedestrian's height, headform impact tests were conducted.

ACCIDENT ANALYSIS

We investigate vehicle-pedestrian impacts using the current accident data. The police and in-depth accident databases of ITARDA (Institute for Traffic Accident Research and Data Analysis) are used for this analysis. The police data are based on accident reports which provide the most voluminous data from all over Japan. The in-depth accident data is based on the investigations of a limited number of accidents collected by the ITARDA. In this database (1993-1997), the number of selected pedestrian accidents is 104.

POLICE DATA ANALYSIS

<u>Car shape and pedestrian injury</u> – From 1992 to 1996, 14,761 pedestrians were killed in traffic accidents in Japan. Among those, pedestrian fatalities, 1041 (7.1%) were struck by mini cars, 6646 (45.0%) by ordinary-sized

cars, 3628 (24.5%) by mini vans, and 965 (6.5%) by large-sized vehicles. In order to examine the injury risks to body regions by vehicle shape, the numbers of fatalities and serious injuries in passenger car accidents were analyzed. Only the most severely injured body regions were considered. Vehicles were divided by class: mini car, small sedan, medium sedan, large sedan, sports and specialty car, wagon, mini van and SUV (Table 1). The mini car is the domestic category in Japan whose size and engine displacement are prescribed in the regulations as: width < 1.48 m, length < 3.4 m, engine displacement < 0.66 L. In this research, small, medium and large sedans are defined based on engine displacement (The engine displacement of the small sedan is less than 1.5 L, that of the medium sedan is from 1.5 to less than 2.0 L, and that of the large sedan is 2.0 L or more). The injury risks are also compared for two classes of vehicles with different shapes, i.e., bonnet-type cars (mini car, small, medium, large sedan, sports and specialty, and wagon) and mini vans.

In order to reduce the influence of impact velocity and variations in pedestrian height, the accident data were limited to a vehicle of traveling velocity of less than 40 km/h and a pedestrian age of 13 years or over. Using this

method, the influence of vehicle shape on pedestrian injuries can be more realistically determined.

Table 1. Car classes. ( ) shows the corresponding models in the US or the EU.

Vehicle class		Vehicle example
Mini car		Alto, Mira, Minica, Wagon R, Life
Small sedan		March (Micra), Corolla, Sunny (Sentra), Civic
Medium sedan		Camry, Bluebird (Altima), Accord, Galant
Large sedan		Mark II, Crown, Celsior (LS 400), Legend (RL)
Sports & Specialty		Celica, 180SX (240SX), Eclipse, RX-7
Wagon		Legacy, Odyssey, RVR, Mark II Wagon
Mini van		Estima (Previa), Townace, Serena, Delica
SUV		Land Cruiser, Pajero (Montero), Jimny, Rav4

The results are shown in Figure 1. The head is the leading body region involved in fatalities for all vehicle classes. The number of fatalities due to head and chest injuries is about double for mini vans compared to

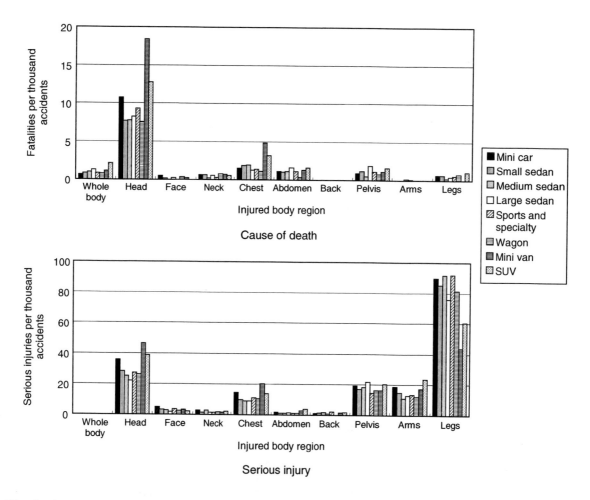

Figure 1.   Distributions of pedestrian injuries per thousand accidents by body region, car shape and injury severity for a traveling velocity of 40 km/h or less, and pedestrian age of 13 or more. The front of the car impacted these pedestrians (1994-1996).

bonnet-type cars. In a serious injury, the number of head and chest injures which can cause death is higher for mini vans than for bonnet-type cars. Therefore, the front of a mini van seems to have more dangerous shape for a pedestrian, whereas, that of a bonnet-type car is more of a threat to the legs as the high number of serious leg injuries confirms.

The risk of head injury to a pedestrian when struck by a mini car is higher than for other bonnet-type cars. The SUV has high aggressivity to the head and chest due to the height of the hood edge and bumper and stiff vehicle body. Since the vehicle front shape affects the distribution of injured body regions, modification of the vehicle shape can be effective for pedestrian protection.

IN-DEPTH ACCIDENT DATA ANALYSIS

Impact velocity and pedestrian injury – The cumulative distributions of the impact velocity for bonnet-type cars and mini vans were compared (Figure 2). When struck by bonnet-type cars, the impact velocity at a cumulative frequency of 50% is about 40 km/h for severe injuries (MAIS 3-6) and 25 km/h for minor injuries (MAIS 1, 2). In the case of a mini van, the velocity for severe injuries is 25 km/h, and that for minor injuries is 15 km/h. The impact velocity of the 50% cumulative frequency of mini vans is much lower than that of bonnet-type cars for both levels of injury. These results confirm that a pedestrian injury is more likely to be severe when struck by a mini van, even when the impact velocity is lower than that of bonnet-type cars.

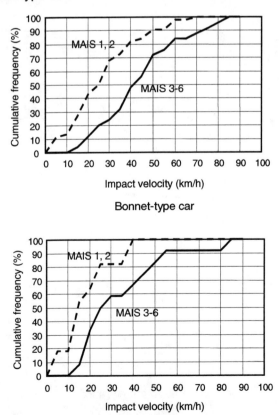

Figure 2. Vehicle shape and impact velocity

Head impact location – Head impact locations on the vehicle were examined for bonnet-type cars and mini vans. This analysis shows the effects of various vehicle front shapes on head impact positions and on the potential head injury risk to the pedestrian. Head injuries sustained by impact with the vehicle body were counted, whereas those by impact with the ground were omitted. Head contact positions are shown in Figure 3. In impacts against bonnet-type cars, the most frequent head impact locations are the lower part of the windscreen and its vicinity; the windscreen frame, cowl and A pillar. Thus, to protect pedestrians, the modifications to those areas of the vehicle should be considered. In addition, such accidents often occurred on the passenger side of the cars. As for impacts with mini vans, the frequency of head contact with the windscreen frame is high.

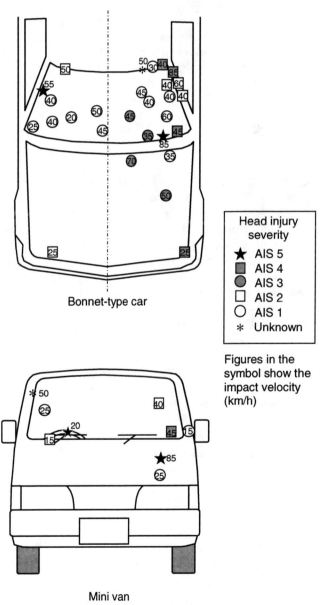

Figure 3. Head impact locations. The results of a bonnet-type car is taken from ITARDA [7]

Wrap around distance (WAD) – A pedestrian hit by a car is wrapped around the front of the vehicle. Therefore, head impact location is estimated by the WAD, which is the length measured along the vehicle's front profile from the ground to the head impact location. The WAD has a close relation to the standing height of the pedestrian. Figure 4 shows the relation between pedestrian height and the WAD for two vehicle types. The WAD increases with pedestrian height, and exceeds that height when the pedestrian is involved in an accident with a bonnet-type car. On the other hand, in the case of a mini van, the WAD approximately equals the pedestrian height.

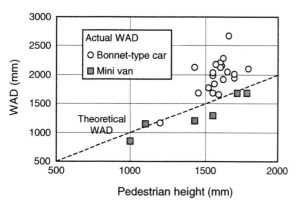

Figure 4.   Pedestrian height and WAD

Head injuries and their causes – The causes of injury in accidents with bonnet-type cars are shown in Figure 5. The number of injuries to the head is counted, and includes those due to the impact against the vehicle body of bonnet-type cars in 9 cases with severe injuries (AIS 3-6), and in 22 cases with minor injuries (AIS 1, 2). The windscreen frame and A pillar have a high potential to cause severe head injury (17%) followed by the hood (11%) and the fender (6%). For minor injuries, the windscreen showed the highest frequency (32%) among the injury causes of all vehicle parts. There are three cases where the windscreen caused severe injuries to the head, and in those three, the impact location was close to the windscreen frame. The A pillar, hood and fender also lead to severe head injuries. Thus, to reduce head injury severity, vehicle body parts such as the A pillar, hood and fender, should be modified. On the other hand, ground impact caused severe injury to the head in 17% of the cases, and appears to be one of the head contact locations with a high frequency. For minor injuries, ground impact was the most frequent cause of injury (41%). Countermeasures against head injuries due to ground impact are difficult to take, which largely limits pedestrian protection to design modifications of the car.

As for mini vans (Figure 6), among the cases of head impact on the vehicle, 10 sustained severe injuries, and 8 had minor injuries. The front panel (42%) and windscreen frame (17%) were the main injury causes for the severe injuries. The windscreen is responsible for the highest frequency of minor injuries. Furthermore, impact with the ground involved two cases (17%) of severe injuries and five cases (38%) of minor injury. For mini vans, modifying the front panel should be considered to improve pedestrian protection.

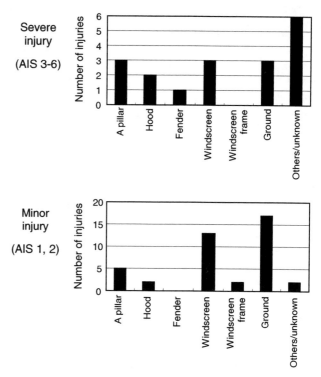

Figure 5.   Causes of head injury (bonnet-type car)

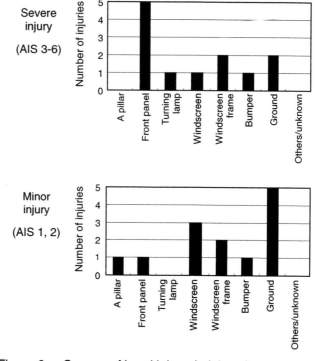

Figure 6.   Causes of head injury (mini van)

Various types of head injuries occur in accidents. Figure 7 shows the distribution of head injuries classified by injury causes, type and severity. Brain injury is a main cause of the severe head injuries (AIS 3-6). On the other hand, for the minor injuries (AIS 1, 2), the percentages of bruises and lacerations are high. Especially, in contact with the windscreen, the pedestrian is likely to sustain bruise and laceration unless impact locations are near the windscreen frame. Skull fractures occur frequently when a pedestrian's head contacts with a stiff part such as an A pillar, windscreen frame or the ground.

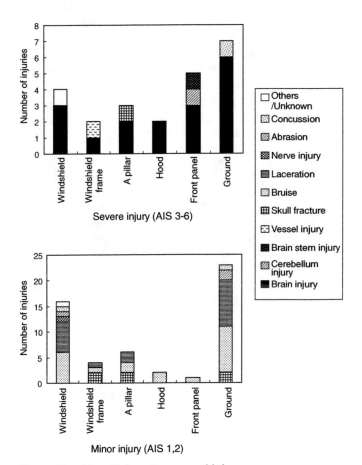

Figure 7.   Head injury types and injury causes

Some typical accidents in which pedestrians sustained injury to the head in contact with different areas of the car body were examined. The three accidents involved bonnet-type cars, and the impact locations were the windscreen, the lower area of the windscreen and the A pillar.

A pedestrian (male, 46-years-old, height unknown) was hit by a bonnet-type car at an impact velocity of 20 km/h. His head impacted the windscreen (Figure 8). Since the crush depth of the windscreen, made of laminated glass, was large (60 mm) there was only minor injury to the head (AIS 1, headache or dizziness). The kinetic energy of the head was absorbed by the large deformation of the windscreen.

A pedestrian (female, 64-years-old, height 152 cm) was struck by a bonnet-type car at an impact velocity of 35 km/h. The head hit the windscreen close to the windscreen frame (Figure 9). The area of the spider-webbed pattern is smaller than when contact is made in the center of the windscreen, as shown in Figure 7. In this accident, the crush depth of the windscreen (laminated glass) was small (10 mm) and the energy absorption of the car at impact was also small. Therefore, this pedestrian sustained serious injury involving a subarachnoid hemorrhage (AIS 3), oculomotor nerve NFS (AIS 2), and lacerations (AIS 1).

A pedestrian (male, 31-years-old, height 162 cm) was hit by a bonnet-type car at a velocity of 55 km/h. As shown in Figure10, the head made contact with the A pillar and the windscreen, and the crush depth of the A pillar was small (5 mm). He sustained diffuse axonal injury (AIS 5), epidural hematoma (AIS 4), and skull fractures (AIS 2).

These accidents demonstrate that the injury patterns of the head depend on the stiffness of the contacted parts of the car. The head injury risk is low when contact is with the windscreen. However, even then, if the impact location is close to the frame or A pillar, serious injuries to the head can occur.

Figure 8.   Head contact with driver side windscreen

Figure 9.   Head contact with lower area of windscreen

Figure 10. Deformation of A pillar by head impact. The head struck both the windscreen and A pillar.

MATHEMATICAL SIMULATION

In order to clarify the relation between injury risk and vehicle shape, mathematical simulations were performed and injury mechanisms are discussed for three types of vehicles, i.e., mini car, medium sedan and mini van.

MODEL DEVELOPMENT – The pedestrian model used in the mathematical simulations is based on the human-body model which was developed by Yang and Kajzer [8]. This is a multi-body simulation model which consists of rigid bodies and joints (Figure 11). Ellipsoids are attached to the rigid body to represent the surface of the body. Similar to the original model, the joint characteristics are based on the human body, and contact interactions between body and car were obtained from the experiments [9].

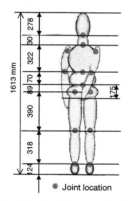

Body segment	Mass (kg)
Head	4.7
Neck	1.1
Chest	14.7
Abdomen	3.4
Pelvis	9.7
Upper arms	3.2
Lower arms	3.6
Upper legs	11.9
Lower legs	5.5
Feet	1.3
Total	59.0

Figure 11. Pedestrian model

In Japan, many elderly people ranging in height from 140 cm to 165 cm are involved in pedestrian accidents [10]. Their average height is lower than that of the AM50. Therefore, a pedestrian model has been developed focusing on an average Japanese male aged 60 to 69 whose standing height is 161.3 cm and weight is 59.0 kg [11]. The sizes, masses and inertia of the body segments of the pedestrian model were calculated using GEBOD (Generator of Body Data).

The vehicle models consist of the bumper, hood edge, hood and windscreen. The bumper, hood edge and hood are represented by ellipsoids, and the windscreen by a plane. Considering that in real-world accidents many pedestrians are struck when crossing the road, in the model the vehicle strikes the side of the pedestrian body. The impact velocity in the simulation is 40 km/h.

SIMULATION RESULTS – The model shows various kinematics upon impact with different vehicle shapes, as shown in Figures 12, 13 and 14. When a pedestrian is struck by a mini car or a medium sedan, the bumper hits the leg and the hood edge hits the thigh, while the upper torso rotates toward the hood of the car. The pedestrian's head impacts the lower region of the windscreen in 97 ms when hit by the mini car, and the hood rear area in 118 ms when hit by the medium sedan. With the mini van, the whole body is struck by the vehicle front at almost the same time. The chest strikes the upper part of the front panel, the head contacts the lower part of the windscreen at 41 ms, while the entire body is projected out ahead of the vehicle.

It was found that the pedestrian kinematics on impacting a vehicle, involve both translational and rotational movements. In the case of bonnet-type cars such as a mini car or a medium sedan, the pedestrian's upper body rotates against the vehicle, and this rotational movement is dominant. Whereas with a mini van, the translational movement is dominant, and the pedestrian is pushed forward in front of the vehicle as shown in Figure 14.

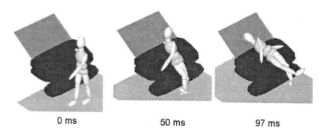

Figure 12. Pedestrian kinematics (mini car)

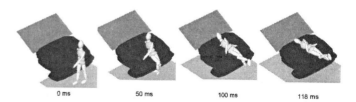

Figure 13. Pedestrian kinematics (medium sedan)

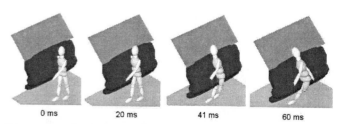

Figure 14. Pedestrian kinematics (mini van)

The head impact velocity affects injury severity. The head's resultant velocity relative to the vehicle is compared among three types of vehicles (Figure 15). Because the translational movement dominates pedestrian kinematics in a mini van impact, the resultant head velocity decreases consistently after impact. The head contacts the vehicle at a velocity of 9.4 m/s, which is lower than the initial velocity. In the case of the medium sedan and the mini car, the influence of both translational and rotational movements is significant. In the first phase, the resultant head velocity increases due to the rotational movement of the upper body, then decreases due to the translational movement of the whole body. The head's resultant velocity on impact is 12.0 m/s for a mini car and 9.6 m/s for a medium sedan. It is clear that the resultant head velocity depends on the shape of the vehicle.

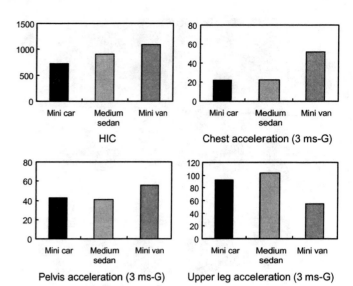

HIC

Chest acceleration (3 ms-G)

Pelvis acceleration (3 ms-G)

Upper leg acceleration (3 ms-G)

Figure 16. Pedestrian injury parameters

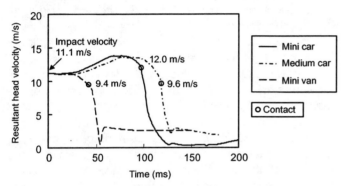

Figure 15. Resultant head velocity to the vehicle

Injury parameters such as the Head Injury Criteria (HIC), and the acceleration of the chest, pelvis and upper leg (3 ms) are shown in Figure 16. In impact with a mini van, the HIC level is highest because the head makes contact with the windscreen frame which is a stiff structure. This finding of a high HIC level in the simulation is consistent with that of our accident analyses showing that injury risk is high when impacted by a mini van. On the other hand, the HIC level is lower upon impact with a mini car because the head makes contact with the windscreen itself which is less stiff.

The threshold of severe chest injury is 90g (TTI) for the chest acceleration in a side impact. As observed in Figure 16, although the chest accelerations in all simulations are less than 90g, the acceleration on impact with a mini van is twice as high as that with other vehicle types. This result also agrees with that of our accident analyses showing the chest injury risk to be higher when struck by a mini van. The pelvis acceleration on impact with a mini van is the highest, though this level is still lower than the injury threshold in a side impact (130g). The acceleration of the upper leg when struck by a mini car or a medium sedan is higher than when struck by a mini van. This result is also consistent with that of our accident analyses.

HEAD IMPACT TEST

METHODOLOGY – The pedestrian's head can make contact with vehicle various areas different in stiffness. Therefore, headform impact tests were carried out to evaluate injury risk to the head on impact with front areas of the car. The adult headform impactor prescribed for the proposed EEVC pedestrian test procedures [12] was used. The outer layer of the impactor is composed of a skin and sphere, with a mass of 4.8 kg. The acceleration is measured at the impactor's center of gravity. The impact velocity is 40 km/h and the impact angle is 65 degrees from the horizontal plane. Various impact locations such as the hood top (WAD is 1500 or more), cowl, fender and the lower area of the windscreen were impacted. In the case of the windscreen, the impact positions varied in proportion to the distance from the windscreen frame and A pillar. The HIC (36 ms) was calculated for impacts on each area of the car. The headform impact tests against the windscreen are shown in Figure 17.

The velocity has a large effect on injury risk to the pedestrian. The results of simulation showed that the pedestrian's head hit the vehicle at differing velocities depending on the vehicle shape (see Figure 15). Therefore, we performed impact tests against the hood and windscreen at impact velocities of 30, 40 and 50 km/h, and compared the HIC values.

The same small car model (Toyota Collora AE91) was used in the tests. The windscreen of this car is of laminated safety glass which consists of three layers: an outer glass layer, a Polywinyl Butyral (PVB) interlayer and an inner glass layer. The thickness of the outer and inner glass is 2.3 mm, and that of the interlayer is 0.76 mm, which are the commonly used specifications for windscreens.

Figure 17. Head impact test on the windscreen

TEST RESULTS

Impact location and injury risk – The impact locations and calculated HICs are shown in Figure 18. A total of 38 impact tests were carried out on the hood, fender, cowl, windscreen, windscreen frame and A pillar. For the hood, cowl and fender areas which are prescribed in the EU test procedures, the HICs for only two locations are less than the injury threshold (HIC 1000). The rear hood and hood/fender areas produce high HICs. The HICs are extremely high (over 5000) for the hood hinge, the hood stopper, the corner of the windscreen frame, and the bottom of the A pillar.

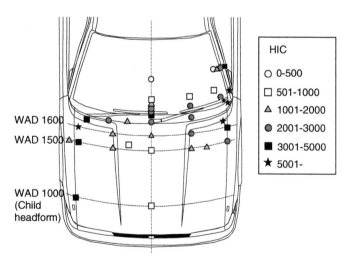

Figure 18. Distributions of HIC and impact location by impact position for the tested car (40 km/h). The impact tests of WAD 1000 were performed by using a child headform impactor.

The car body shows various force-deformation characteristics when hit by the headform. Figure 19 shows the force-deformation characteristics of the main locations of the car. In the hood region, the force reaches

a peak deformation of 25 mm, and after the hood reinforcement separates from the hood, the force levels off. However, the hood at the hinge and the hood stopper produce high force levels of 20 kN. In the cowl area, the force increases consistently, whereas at the wiper pivot, the force is high due to the deformation of the wiper pivot axis. The A pillar has a constant force level due to the collapse of its box shape, but its force level is high enough to cause serious injuries to the head.

In addition to the baseline force-deformation characteristics of each car body parts, it was found that the local high stiffness of the hood hinge, hood stopper and wiper pivot has a major effect on both the force-deformation characteristics and the HIC.

The force-deformation characteristics were compared among the lower edge of the windscreen frame, 50, 150 mm above it, as well as at the windscreen center (Figure 20). In the windscreen area 50 mm above the lower windscreen frame, the force shows an inertial spike of about 7.5 kN in the initial phase when the glass breaks. After that, the force increases, and the force-deformation curve is similar to that of the windscreen frame. For the impact on the center of the windscreen, the initial spike of the glass breaking is followed by a low plateau force of about 3 kN, which is due to stretching of the PVB interlayer of the HPR glass. In this area, the effect of the stiffness of the lower windscreen frame on the force-characteristics is small. These results show that the force-deformation characteristics of the windscreen are mainly affected by those of windscreen frame.

The relation between the HIC and the distance from the windscreen boundary is examined along three paths as shown in Figure 21. The HIC value is at a maximum at the windscreen boundary for all paths, and it decreases with the distance from the boundary. The inclination toward decrease in the HIC varies with each boundary. The HIC value of path A decreases gradually with the distance from the lower windscreen frame because the headform impactor interferes with the top of the instrument panel. However, for the A pillar, the HIC decreases abruptly (path B). Since the windscreen's bonded width at the A pillar is small, the deformation of the windscreen increases in such a way that the windscreen boundary of the A pillar works like a hinged joint. The corner of the windscreen boundary is so stiff that the HIC in the windscreen around this corner reaches a high value (path C). The HIC of path C shows a similar tendency to that of path A when the distance from the lower windscreen frame is over 100 mm, which means that the influence of the boundary by the A pillar is small in this region.

A contour map of the windscreen is drawn based on the test results, as shown in Figure 22. The region where the HIC value is below the injury threshold occupies a large proportion of the windscreen.

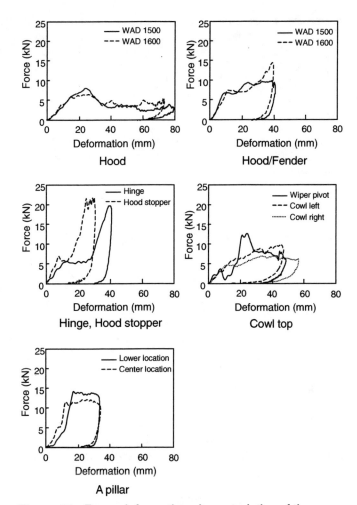

Figure 19. Force-deformation characteristics of the car from headform impact tests (40 km/h)

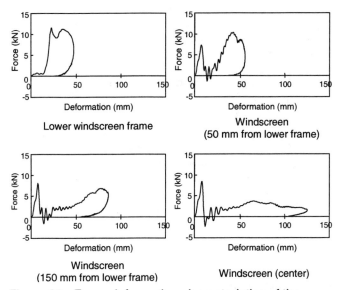

Figure 20. Force-deformation characteristics of the windscreen from headform impact tests (40 km/h)

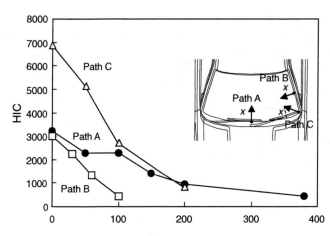

Distance from the windscreen boundary, x (mm)

Figure 21. The relation between the HIC and the distance from the windscreen boundary of the tested car (40 km/h). The path A is from the lower windscreen frame, the path B is from the A pillar, and the path C is from the corner of the windscreen. For path C, the lateral axis indicates the distance from the lower windscreen frame.

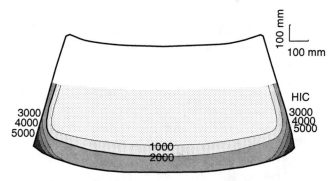

Figure 22. HIC in the windscreen region in the headform impact tests for the tested car (40 km/h)

Impact velocity and injury risk – In order to clarify the effects of impact velocity, its relation to the HICs was examined for the hood (WAD 1500 mm on the centerline of the car) and the center of the windscreen. The results are shown in Figure 23. The hood produces a linear increase in the HIC with increasing impact velocity, and the HIC value exceeds 1000 at 50 km/h. When the impact velocity is 50 km/h on the windscreen, the interlayer of the windscreen was torn (there was no penetration of the headform), which results in a HIC value below the injury threshold. The fragments of broken glass become larger with higher impact velocity. Since the HIC for impact with the windscreen is still less than the injury threshold even at the impact velocity of 50 km/h, it is considered the injury risk to the head is low in the center of the windscreen.

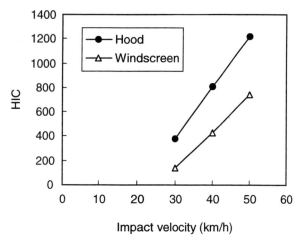

Figure 23. Effect of the impact velocity on the HIC for the tested car

HIC AND DYNAMIC DEFORMATION – The deformation necessary to keep the HIC below 1000 is important in order that a car may be designed to reduce the likelihood of pedestrian head injuries. MacLaughlin et al. [13] showed in headform impact tests onto the hood top (37 km/h) that the HIC is related to the dynamic deformation. Since their study experimentally investigated only the hood top, we examined this relation based on theoretical analysis as well as on impact tests for the windscreen and the bonnet top.

Theoretical HIC – We examined the relation between the HIC and dynamic deformation based on the approximation of the acceleration curve. Let the deceleration-time history of the headform $\alpha(t)$ [m/s^2] be approximated based on the curve of the second degree of time t (see Figure 24):

$$\alpha(t) = -a\, t\,(t - 2t_0) \tag{1}$$

where a is the coefficient for curve fitting, and t_0 is the time when the deceleration reaches maximum.

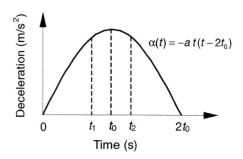

Figure 24. Model of headform deceleration

According to its definition, the HIC is calculated using deceleration as follows:

$$HIC = \max_{0 \le t_1 \le t_2 \le 2t_0}\left[\left\{\frac{1}{t_2 - t_1}\int_{t_1}^{t_2}\frac{\alpha(t)}{g}\,dt\right\}^{2.5}(t_2 - t_1)\right] \tag{2}$$

where g is the acceleration of gravity (=9.81 m/s^2). We assume that t_1 and t_2 are symmetry with respect to t_0 ($t_0=(t_1+t_2)/2$). By introducing the variable $\tau=t-t_0$, the HIC can be computed as follows:

$$HIC = \max_{0 \le t_1 \le t_0}\left[\left\{\frac{1}{t_1}\int_0^{t_1}\frac{a(\tau^2 - t_0^2)}{g}\,d\tau\right\}^{2.5}(2t_1)\right] \tag{3}$$

$$= \sqrt{2}\left(\frac{5a}{6g}\right)^{2.5} t_0^6 \tag{4}$$

The relation between the HIC and dynamic deformation x_d (m) is obtained as:

$$HIC = 2^{-9}3^{1.5}5^{2.5}\,g^{-2.5}v_0^4\,x_d^{-1.5} = 0.001882\,v_0^4\,x_d^{-1.5} \tag{5}$$

where v_0 is the initial velocity (m/s). We call this calculated value the theoretical HIC, which increases markedly with velocity (v_0), and decreases with dynamic deformation (x_d). The HIC value is below 1000 if

$$x_d > 0.0934 \text{ m} \tag{6}$$

Based on this analysis, a dynamic deformation of 93.4 mm can make the HIC value less than the injury threshold.

Experimental results – The HIC results obtained from the headform impact test on the car body (excluding the windscreen) and the windscreen are shown as a function of dynamic deformation in Figure 25. The HIC correlates well with the dynamic deformation of the car body and windscreen. The HIC of the windscreen is larger than that of the car body. This tendency is apparent for HIC values below 3000. It can be considered that the high HIC of the windscreen is due to the inertial spike of acceleration in the initial phase. Figure 25 also shows that the theoretical HIC calculated using Eq. (5) agrees with the headform test results.

The approximation curves were calculated for the windscreen and the car body. Based on these approximation curves, a HIC value of 1000 is associated with a dynamic deformation value of 76 mm for the car body, and 89 mm for the windscreen. In order to reduce the HIC below 1000, dynamic deformations greater than those values are necessary.

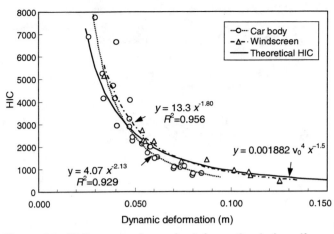

Figure 25. HIC versus dynamic deformation in headform impact tests for the tested car (40 km/h)

DISCUSSION

The risk of head injury depends on the impact velocity of the head and the stiffness of the vehicle structure where the head makes contact. Furthermore, the mathematical simulation showed that the head impact velocity and position are affected by the vehicle shape since the pedestrian kinematics depend on it. Therefore, the vehicle front shape should be considered for reducing head injury risk to the pedestrian.

Based on the in-depth accident data, the head impact locations have shifted from the car front towards the windscreen area because of the short and steep bonnets of modern cars. The headform impact tests were also carried out for various car areas including the windscreen. The windscreen center produces a low HIC even at a high velocity of 50 km/h, whereas the HIC value increases for the windscreen close to the frame. Thus, to reduce the likelihood and severity of head injury, it would be effective to design a vehicle shape considering pedestrian kinematics in order that the head would contact the windscreen center.

In determining the head impact test procedure for the windscreen, it is sufficient to test only near the boundary of the windscreen, and not the center area. This is because the HIC reaches a maximum at the boundary of the windscreen. It is also necessary to take into consideration that the HIC has its highest value at the lower corner of the windscreen boundary.

The dynamic deformation required for a HIC value below 1000 is 76 mm for the car body, and 89 mm for the windscreen. However, it will be difficult to ensure this deformation for stiff parts of the car such as the A pillar, cowl and windscreen boundary. One feasible solution would be to design a car so that the pedestrian's head does not contact these areas on impact. For example, the cowl and windscreen frame could be covered by the hood panel, and the A pillar be located to the side and rear of their location in the modern car.

Accident analysis using police data indicated that in impact with a mini car, the head injury risk to the pedestrian is high (see Figure 1). This high head injury risk could not be reproduced by mathematical simulation because the head made contact with the windscreen in the simulation. The headform impact tests showed that the HIC values are high near the windscreen boundary. Since the mini car has a small windscreen area, the pedestrian may incur a higher risk of contact with the windscreen boundary. In particular, the heads of shorter pedestrians can readily make contact with the lower windscreen region. This could be one of the reasons for high head injury risk in accidents involving mini cars.

CONCLUSIONS

Head injury risk on pedestrian impact with vehicles was examined based on accident analyses, mathematical simulation, and headform impact tests.

From accident analysis, it was found that when pedestrians are struck by bonnet-type cars, they tend to sustain serious injuries to their legs. In impact with mini vans, they are at high risks of serious and fatal injury to their head and thorax. The probability of severe pedestrian injuries is higher for a mini van than for a bonnet-type car. This is because, in mini-van impacts, the head of the pedestrian hits a stiff area such as the windshield frame or A pillar. The windscreen center, on the other hand, seldom causes serious injury to the pedestrians head.

The mathematical simulation showed that the acceleration of the upper leg was found to be high in a bonnet-type car impact, while that of the chest and the HIC is high in a mini van impact. These results are consistent with our accident analyses of body regions likely to be injured by these vehicle types.

From the headform impact tests we obtained the distribution of the HIC value in the windscreen and found that the HIC is maximum at the boundary of the windscreen. Based on the relation between the HIC and dynamic deformation, such deformations greater than 89 mm are necessary for the windscreen to reduce the HIC below 1000.

REFERENCES

1. Cavallero et al.: Improvement of Pedestrian Safety: Influence of Shape of Passenger Car-Front Structures Upon Pedestrian Kinematics and Injuries", SAE Paper 830624, 1983.

2. Ishikawa, H., Yamazaki, K., Ono, K. Sasaki, A., "Current Situation of Pedestrian Accidents and Research into Pedestrian Protection in Japan", Proceedings of 13th ESV, pp. 281-293, 1991.

3. Higuchi, K., Akiyama, A., "The Effect of the Vehicle Structure's Characteristics on Pedestrian Behavior", 13th ESV, 1991.

4. EEVC, "Improved Test Methods to Evaluate Pedestrian Protection Afforded by Passenger Cars", EEVC Working Group 17 Report, 1998.

5. Harris, J., "A Study of Test Methods to Evaluate Pedestrian Protection for Cars", Proceedings of 12th ESV, pp. 1217-1225, 1989.

6. NHTSA, "Pedestrian Injury Reduction Research", Report to the Congress, 1993.

7. ITARDA, Report of In-Depth Accident Analysis, 1996 Edition 1997 (in Japanese).

8. Yang. J., Kajzer, J., Computer Simulation of Impact Response of the Human Knee Joint in Car-Pedestrian Accidents, SAE Paper 922525, 1992.

9. Ishikawa, H., Kajzer, J., Schroeder, G., "Computer Simulation of Impact Response of the Human Body in Car-Pedestrian Accidents", SAE Paper 933129,1993.

10. Yoshida, S., Matsuhasi, T., "Development of Vehicle Structure with Protective Features for Pedestrians", SAE Paper 1999-01-0075, 1999.

11. Management and Coordination Agency of Japan, Statistics Bureau and Statistics Center, Japan Statistics, 1997 (in Japanese).

12. EEVC, "Proposal for Methods to Evaluate Pedestrian Protection for Passenger Cars", EEVC Working Group 10 Report, 1994.

13. MacLaughlin, T. F., Kessler, J. W., "Test Procedure – Pedestrian Head Impact Against Central Hood", SAE Paper No. 902315, Society of Automotive Engineers, 1990.

2003-01-1300

Body Concept Design for Pedestrian Head Impact

S. Iskander Farooq and Peter J. Schuster
Ford Motor Company

ABSTRACT

In 1996, the European Enhanced Vehicle Safety Committee, Working Group 17 (EEVC WG17) proposed a set of impact procedures to evaluate the pedestrian injury risk of vehicle fronts. These procedures address three aspects of pedestrian protection – head impacts, lower limb impacts, and thigh impacts – through vehicle subsystem tests. The criteria assessed during these impact tests are affected by the design of most parts of the vehicle body front-end.

One of the challenges to vehicle design introduced by these tests is the impact of an adult pedestrian headform to the top of the fender. The proposed acceptance level for Head Injury Criterion (HIC) is less than 1000 during impacts at 40 km/h. This paper uses the finite element (FE) method to predict the influence of proposed fender and shotgun design modifications aimed at meeting this target. In addition, the known issues with the implementation of these proposed changes are discussed.

Although the proposed changes are shown to meet the target in the theoretical analyses presented in this paper, these changes are also demonstrated to conflict with other aspects of vehicle safety (frontal visibility and frontal high-speed impact), vehicle manufacturing, and durability.

INTRODUCTION

BACKGROUND

Pedestrian injuries can be distributed over the body (Table 1), but the majority of fatalities are due to head injuries (Table 2). Table 3 identifies the locations of pedestrian contact associated with head injuries. Analysis of the data summarized in these tables indicates that the greatest benefit in reducing head injuries—and hence the key areas for improvement—will come from addressing:

1. Head impact to windshield & surrounding areas
2. Head impact to ground
3. Head impact to hood/fender top

Table 1: Distribution of Pedestrian AIS 2+ Injuries [1]

Head/Neck	28%
Chest/Abdomen/Pelvis	22%
Lower Extremities	31%
Upper Extremities	19%

Table 2: Causes of Pedestrian Fatalities [1]

Head Injury	62%
Neck Injury	10%
Spinal Column	3%
Chest Injury	3%
Blood Loss	3%
Internal Injury	5%
Other	14%

Table 3: Contact surfaces associated with pedestrian head injuries [1]

Contact Surface	
Vehicle Hood & Fender Top	6-16%
Vehicle Windshield & Frame	35-51%
Ground	22-49%
Other / Unknown	10%

EEVC WG17 TEST PROTOCOL

For adult and child headform impacts, the EEVC WG17 test protocol proposes a maximum HIC of 1000 during a

40-km/h impact. The head impact zones are defined by boundaries in the vehicle's longitudinal (X) and transverse (Y) directions. The boundaries of the adult and child head impact zones in the Y direction and X direction are illustrated in Figure 1 and Figure 2, respectively. These zones are bounded in the X direction by "Wrap Around Distance" (WAD) lines created by wrapping a tape measure from the ground onto the vehicle in successive X-Z planes. In the Y direction, the hood/fender ("bonnet top") "Side Reference Line" (SRL) marks the lateral extent of the impact zone. This line is defined as the geometric trace of the highest point of contact between a straight edge 700 mm long and the side of a bonnet, when the straight edge, held parallel to the car's Y-Z plane and inclined inwards by 45°, is traversed down the side of the bonnet top [2].

Figure 1 also illustrates how the most outboard location for adult or child head impact is defined. It is measured from the SRL in the Y direction 65 mm for child head and 82.5 mm for the adult head. This method is intended to ensure that the complete headform is inside the SRL at the start of the impact.

DETAILS OF THE STUDY

Prior work has indicated that a HIC of 1000 can be achieved when the allowable intrusion of the child and the adult head into the underhood package space is at a minimum 65 mm and 80 mm, respectively. Beyond the obvious packaging concerns this clearance introduces a number of component design challenges. One key area of challenge is along the top of the fender at the hood shutline.

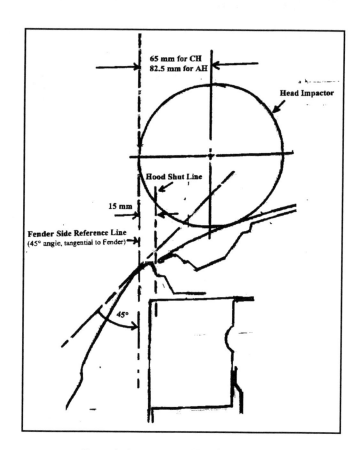

Figure 1: Impact zone side reference line

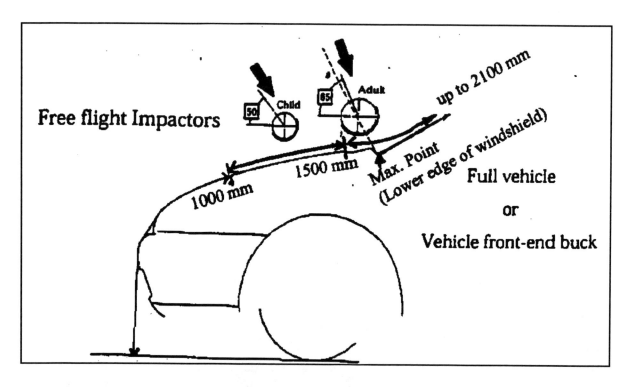

Figure 2: EEVC WG17 head impact zones

68

The first step in this study was to propose designs for the attachment of a fender to a shotgun that allow for the target package space. Figure 3 and Figure 4 show the assembled design proposal, while Figure 5 shows the individual components:

1. A lowered apron structure (consisting of shotgun inner and outer) to provide the required crushable space.
2. A channel bracket to bridge the gap between the fender catwalk and the lowered shotgun. This part also acts as the mounting surface for the fender catwalk. The design of this part makes it crushable under the load of the head impact. This part is welded to the shotgun and the fender is bolted to it. Figure 6 shows different proposals for this part.
3. A crushable fender catwalk design. Unlike the traditional fender catwalk design, which has a stiff, almost vertical downflange, the new concept introduces a step in the down-flange while accommodating the requirement for clearances to any hood overslam event.

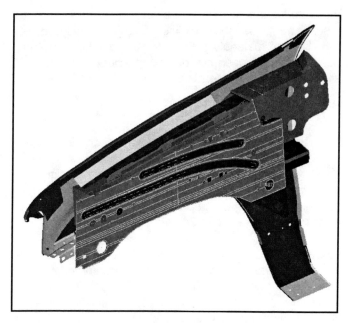

Figure 4: Assembled view of proposed system

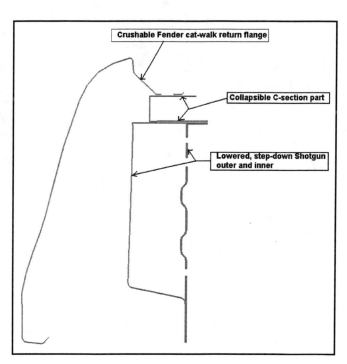

Figure 3: Cross-section through proposed system

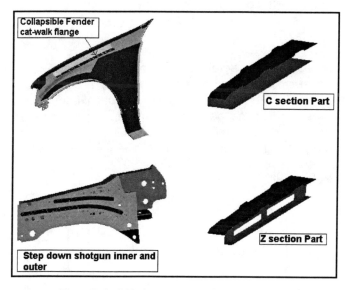

Figure 5: Individual components of proposed system

RESULTS

HOOD AND FENDER MATERIAL

Child and adult headform impact simulations were performed with the target point located directly in the hood-fender gap. Impacts to the baseline design were simulated to determine the current anticipated performance and to establish the effect of material type in the hood and fender. Figure 7 illustrates this impact simulation with the baseline design and Figure 8 shows the deceleration curves of the HIC values for these two

Starting with the baseline performance (without a lowered shotgun, bracket, or crushable catwalk), the system design was changed in the following ways. Each step was compared to determine its effect on the HIC:

1. Hood and fender material properties
2. Channel section and gage
3. Channel height (package space)

materials – namely steel and aluminum. Table 4 lists the results found when the baseline design was analyzed with two different materials for the hood and fender. In this design, the HIC does not differ appreciably for steel and aluminum hoods and fenders.

CHANNEL TYPE AND GAGE

The baseline analyses indicated that changes were necessary to the fender structure and underlying package depth. To improve the results, the design was modified to introduce a "dog-leg" to allow the fender catwalk to crush under impact, and the shotgun structure height was decreased by 24 mm.

Many times, due to formability issues, it is not feasible to design a fender catwalk that can accommodate this lowered shotgun and still be bolted to it. In order to overcome this material formability feasibility issue, a bridging or filler part was added in between. Different variations of this part were analyzed, mainly varying thickness of C-channel, Z-channel and Hat channel type sections. This part also acts as a mounting structure for the fender onto the body. Figure 4, Figure 5, and Figure 6 show the proposed design of the components of the new closure system. CRLC steel properties were used for these channel parts.

The headform deceleration analysis results with different channel cross-sections and material gages are shown in Figure 9 and Figure 10. As can be seen from these curves, a C-channel part with 0.8 mm gage provides the lowest peak deceleration for the adult headform. The same part with 1.0 mm gage results in the lowest peak deceleration for the child headform. The corresponding calculated HIC values are listed in Table 5.

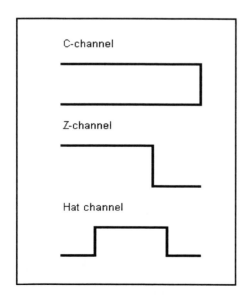

Figure 6: Channel cross-sections

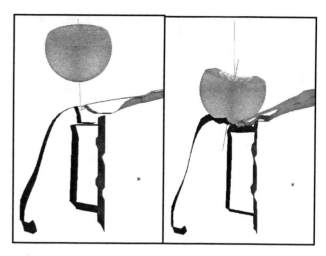

Figure 7: Headform impact – baseline design

Table 4: Influence of material on HIC (baseline design)

Material	HIC	
	Adult	Child
Steel (0.7 mm, CRLC)	1720	2422
Aluminum (0.9 mm)	1646	2341

Table 5: Influence of channel type and gage on HIC

Component	Gage	HIC	
		Adult	Child
C-channel	0.8	980	970
	1.0	991	950
	1.2	1007	1007
Z-channel	1.0	1071	1024
Hat-channel	1.0	1222	n/a

PACKAGE SPACE

To determine the influence of package space alone on the headform decelerations, the shotgun was lowered further, from 24 mm to 33 mm. These changes were implemented by increasing the height of the vertical wall of the 0.8 mm gage C-channel section. Simulations were performed only for the adult headform. Figure 11 shows the deceleration plots while Table 6 summarizes the calculated HIC from these analyses.

Table 6: Influence of package space on HIC (C-channel)

C-channel height (mm)	HIC (Adult)
0 (no channel)	1646
24	980
27	900
30	860
33	815

Table 6: Influence of package space on HIC (C-channel)

DISCUSSION

HEADFORM SIMULATIONS

Hood and Fender Materials

Table 4 indicates that with the baseline design, material type has little influence on the predicted HIC. Both curves in Figure 8 show a very sharp rise in the headform deceleration to an early peak. This early peak is due to overall stiffness of the fender catwalk and the close proximity of the shotgun structure to the outer surface. Although the HIC results are similar, the aluminum deceleration curves show a lower peak value than the steel, likely due to greater deformability of the fender catwalk with this material. The remaining

simulations were performed with aluminum properties to take advantage of this trend.

Channel Section and Gage

In the second step of the system modifications, a fender 'dog-leg' and a 24 mm reduced-height shotgun (with corresponding channel sections to bridge the gap) were introduced in the FE model. These changes resulted in significant improvements in the predicted headform HIC, as seen by comparing Table 4 and Table 5. The differences are also obvious in Figure 9 and Figure 10.

For adult headform impact simulations (Figure 9), the initial jerk of all the curves is the same, but the system modifications allow a delay in the peak deceleration until approximately 15 milliseconds (versus 7 milliseconds for the baseline). This delay is due to the increased package depth. The headform is allowed to translate further before reaching a stiffer component, allowing more energy to be absorbed before the peak acceleration is achieved. This results in a narrower peak, reducing the predicted HIC.

For child headform impact simulations (Figure 10) the initial jerk is also the same, but the overall peak drops significantly with the system modifications. This is because the increased package space has allowed the child headform to be fully decelerated before any contact with stiffer structures, resulting in a lower predicted HIC.

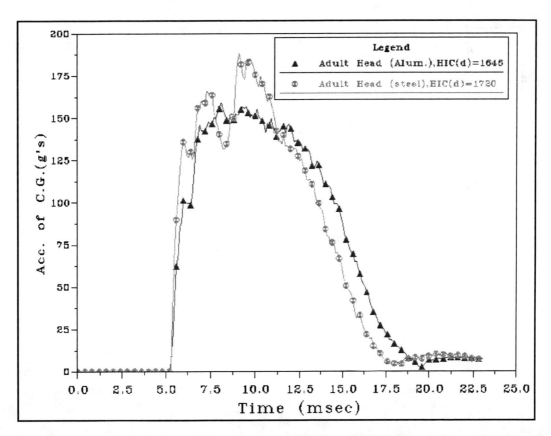

Figure 8: Influence of material properties on headform deceleration (baseline design)

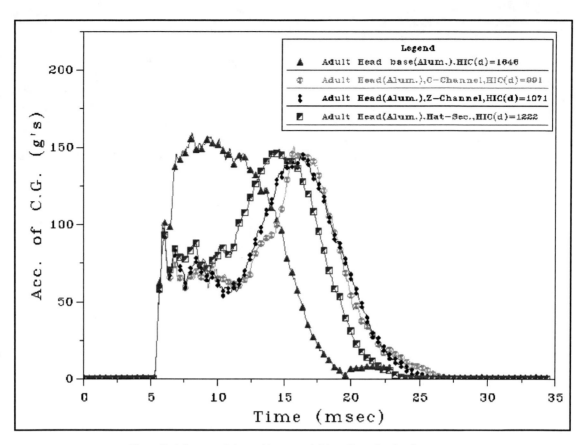

Figure 9: Influence of channel type on adult headform deceleration

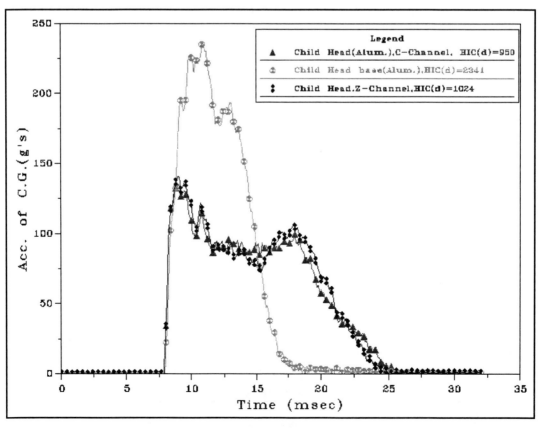

Figure 10: Influence of channel type on child headform deceleration

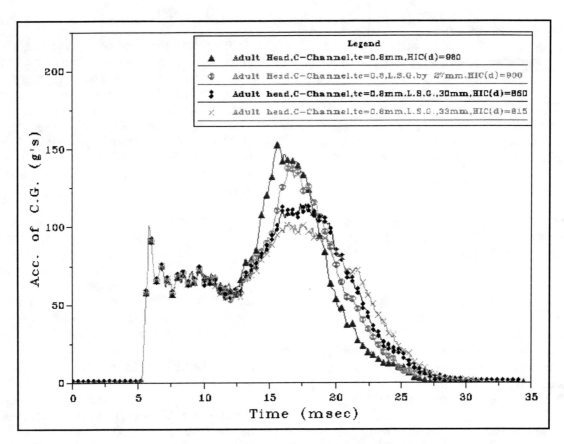

Figure 11: Influence of package space on adult headform deceleration

Table 5 also shows that the shape and gage of the channel section have an influence on the predicted headform HIC. This is more noticeable with the adult headform impact simulations (Figure 9), where the difference in width of the deceleration peaks is due to stack-up of the channel sections near the end of the event. Obviously a design resulting in less stack-up allows more distance to be traveled before greater stiffness is encountered. Of the designs evaluated, a C-channel part with 0.8 mm gage provides this minimum stack-up.

Package Space

By increasing the height of the 0.8 mm C-channel design (while decreasing the height of the shotgun), the overall crushable space was increased. Figure 11 indicates that increasing the section height from 24 to 33 mm reduces the second peak in the adult headform deceleration. These changes are mirrored in the reduced predicted HIC listed in Table 6. From these results it can be seen that a 33 mm high C-section (with correspondingly lowered shotgun) can achieve a robust HIC<1000 result.

OTHER CONSIDERATIONS

It is important to note that this paper reports on a theoretical study. A practicable vehicle structure meeting the target criteria may not be feasible. Many issues need to be resolved to enable the results of this study to be ready for implementation. This section details some of the key issues anticipated with incorporating these changes.

Safety

One of the major changes proposed by this study is increasing the package space between the top of the fender and the shotgun. To physically achieve this result on a vehicle, either the upper shotgun must be lowered or the fender top surface must be raised. Both of these changes compete with other vehicle safety attributes.

Because of the tire clearance envelope on modern vehicles, the lower part of the shotgun is already as low as possible. If the upper shotgun surface is lowered, the net section height of the shotgun will be reduced. The shotgun is one of the primary load-carrying members for vehicle frontal impact, absorbing impact energy and reducing occupant compartment intrusion. A reduction in

73

the section height of the shotgun conflicts with the frontal impact energy management targets.

If, instead, the fender top surface is raised, this will result in increased height along the entire length of the fender. A change of this nature would conflict with many modern vehicles' forward vision targets. Reducing the forward vision on a vehicle unfortunately may lead to an increase in the number of accidents, since the driver's ability to see and respond to external objects is reduced.

Durability

Current vehicle designs include a rather stiff structure just below the fender catwalk for improved vehicle durability. This improvement manifests itself in at least two different ways:

- Palm Loading – Vehicle users often will lean or sit on the edge of the vehicle's front. To support this loading, structure is needed.
- Low Speed Impacts – Motion of the bumper cover and surrounding components during a low speed impact can apply loads to the fender mounts. These loads are resisted by the attachment structure.

Meeting vehicle targets under these vehicle durability loading conditions will be much more difficult with the reduced stiffness design proposed in this study.

Manufacturing

Although a channel section was added partly in response to the known manufacturing issue with a long fender catwalk, several other key manufacturing issues were not investigated in this study.

- Stamping Feasibility issue with the added C-section channel part requires further study. The reduced 0.8 mm thickness and the need for draw depth of 30 mm or higher, may create material formability or metal flow concerns in the manufacturing of this part.

- Dimensional variability due to added part may also create fit and finish or craftsmanship concerns with the exterior surfaces of fender, hood and door outer panels. Traditionally, fenders are attached to shotgun outer along with the hood hinge on the same plane, but this new study will require added channel part onto which fenders will be attached. New dimensional control scheme and new sets of master control surfaces (GDT methodology) will be required to be developed prior to incorporating this proposal in the vehicle assembly process.

CONCLUSION

A theoretical study of pedestrian headform impacts to proposed vehicle front structures has been performed. This series of finite element simulations predict that a lower HIC may be achieved during pedestrian headform impacts by (a) modifications to the fender catwalk area for better deformability, (b) lowered fender apron (shotgun) structure, and (c) a bridging C-channel part in between the fender and the shotgun. Whether a physical structure can be designed to meet these criteria requires further investigation.

In addition, the key issues preventing near-term application of these changes are discussed. In particular, competing performance targets with frontal impact energy management, forward vision, vehicle durability, and manufacturing issues are identified for future work. The present study did not address these concerns, which must be resolved before this design could be considered in future vehicles.

ACKNOWLEDGMENTS

The authors wish to gratefully acknowledge the finite element analysis support of our colleagues in the Ford Safety CAE group. Our sincere thanks go to Thiag Subbian, Jehad Abbas, and Ahmer Aleem for their effort in performing all of the simulations during this project. In addition to their finite element analysis skills, their insight and experience in the energy management of body components were invaluable to the success of this project.

Additional thanks go to Raimondo Sferco and Paul Fay of the Ford Safety Data Analysis group for the accident data analyses.

REFERENCES

1. Data from the *International Road Traffic and Accident Database* (IRTAD) and the *German In-Depth Accident Study* (GIDAS) made available through Ford Motor Company data access, 2002.
2. EEVC, "EEVC Working Group 17 Report: Improved Test Methods to Evaluate Pedestrian Protection Afforded by Passenger Cars." http://www.eevc.org, 1998.

The Development of Procedures and Equipment for European Pedestrian Impact Protection Requirements

Helen A. (Rychlewski) Kaleto and Michael J. Worthington

MGA Research Corp.

ABSTRACT

Over the recent years, tests have been developed which will result in higher levels of protection afforded to pedestrians involved in vehicle impacts. These requirements will result in improved vehicles, which are designed to be safer and less harmful to pedestrians. The requirements include a series of impact tests into the front end of a vehicle with a variety of instrumented forms including an adult headform, a child headform, as well as an upper legform and a complete legform. Although the requirements have not been finalized, it is expected that they are to become part of the European homologation test series soon. Automobile manufacturers that sell cars in Europe are beginning to incorporate the proposed regulations into new car designs. Readers of this paper will gain a much broader understanding of future changes to vehicles in regard to these requirements, the equipment necessary to conduct these tests, and the data which is used to evaluate front end designs for pedestrian protection.

INTRODUCTION

Pedestrian safety is a very important issue in many overseas countries. This is largely due to the enormous number of pedestrians and cyclists found in the urban cities of most European and Asian countries. The Working Group (WG)-10 report from 1994 [1] was published by the European Enhanced Vehicle-safety Commission (EEVC). It detailed the initial proposed directive aimed at providing safety to pedestrians in case of injury due to a collision with a vehicle. This report was updated and enhanced further in December 1998 with the release of the WG-17 report entitled, "Improved test methods to evaluate pedestrian protection afforded by passenger cars"[2]. The WG-10 report consisted of the outline of new proposed tests to enhance pedestrian safety. It detailed four tests considered to cover all of the possible impact scenarios concerning pedestrians. The

report also detailed at length the construction and criteria for each impact device and test procedure. The purpose of the WG-17 report was to analyze the test methods proposed by the WG-10 report and, with respect to new field data, offer suggestions and/or possible adjustments that should be taken into account.

A large effort has been put forth in creating this proposed regulation. The key element is that the proposed changes and additional requirements will reduce the number of fatalities and injuries that occur in pedestrian-vehicle accident scenarios. Based upon the previously mentioned reports, more than 7,000 pedestrians and 2,000 cyclists are killed every year in road accidents in Europe [1]. Also under consideration are the thousands of non-life threatening, yet serious, injuries that occur as a result of vehicle-to-pedestrian impacts.

PEDESTRIAN IMPACT TESTING

Both the WG-10 and WG-17 reports detail the proposed equipment to be used to conduct pedestrian impact testing. The tests themselves are quite unique to any other current safety test standard or regulation. In fact, the WG-10 proposed tests are currently being used in the European New Car Assessment Program (Euro-NCAP). Similar to the US-NCAP, the results of these tests are published to provide consumers with supplemental vehicle safety information. Automakers are not legally required to meet any pedestrian impact requirements at this time, but will have to at some point in the future in order to sell vehicles in Europe.

The proposed tests are broken down as follows:

- Free flight complete legform-to-bumper
- Upper legform-to-bumper (for higher bumpers)
- Guided upper legform-to-bonnet leading edge
- Free flight adult & child headform-to-bonnet top

The difficulty in developing this regulation has been finalizing the impact devices as well as the test methodologies. The WG-17 report recommended the final designs for the impact devices and other equipment concerning the testing including:

- Free flight complete legform
- Guided upper legform
- Free flight adult headform
- Free flight child headform
- Propulsion system
- Data Acquisition System (DAS)
- Calibration requirements and method

A key factor in the development of an appropriate propulsion test system was that similar equipment already exists for dynamic impact testing. Although initially intended for the purposes of Free Motion Headform (FMH) head impact testing, the pneumatically-controlled impactor systems currently used to conduct Federal Motor Vehicle Safety Standard (FMVSS) 201U – Upper Interior Occupant Impact Protection [3] testing can be used to propel the pedestrian headform impact devices as well. FMVSS 201U simulates the vehicle occupant striking the upper interior portions of the vehicle including the pillars, side rails, headers, and roof. Essentially the same type of test will be conducted using the free flight child and adult headforms here except that an exterior portion of the vehicle is tested rather than an interior component. Most current head impact test systems can be adapted or modified to conduct the proposed free flight adult and child head impact pedestrian tests. Issues related to using current equipment include fabrication of adaptor plates, an increase in the required propulsion pressure, and the addition of several data acquisition channels.

OVERVIEW OF DYNAMIC IMPACT TEST SYSTEM – Figure 1 presents a diagram of a current Pedestrian Impact Test System.

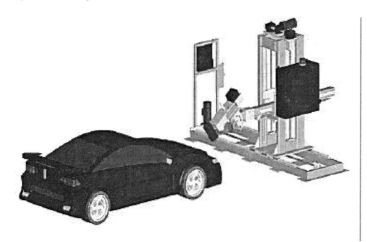

Figure 1. Typical pedestrian system layout.

The test frame shown here provides adjustment in three orthogonal directions. Turn-screw mechanisms are powered by electrical brake motors and are controlled through a hand pendant. The frame arm provides mounting for both pedestrian impactors. Pneumatically charged impactors, using compressed nitrogen gas, provide the acceleration and proper guidance into the test article. A single impactor can be used to conduct each of the legform tests and a second impactor is required to conduct both headform tests. When the system is fully charged and the Data Acquisition System (DAS) is enabled, the impact form is accelerated into the test article and the data is collected and analyzed.

The most significant design factor was compatibility with existing test frames. The complete legform test is similar to an FMVSS 203 BLAK "Tuffy" Dummy test [4] where a sizeable mass is propelled in free flight. The upper legform test resembles an FMVSS 201 instrument panel test where the mass is guided into the test article. As for the headform tests, the biggest difference in comparison to a typical FMH test is that the impact occurs on the exterior of the vehicle. This requires a different fixturing procedure for the test article, but uses essentially the same propulsion system as in the FMH test.

INSTRUMENTED IMPACT FORMS – Figures 2-5 are photographs of the impact forms as proposed in the WG-17 report as well as the propulsion devices.

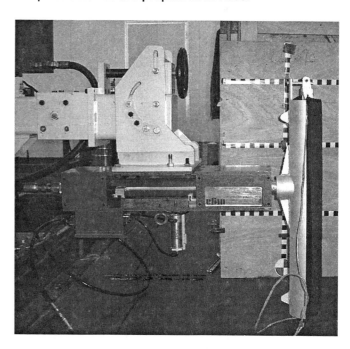

Figure 2. Complete Legform (WG-17 version).

The complete legform is attached to a guide bar, which is attached to the propulsion device. The guide bar is equipped with four pin locator holes that the complete legform is indexed into so that the impact form is kept vertical (must stay within 2° of vertical) during free flight. When the device reaches the required velocity, the complete legform is released from the propelling device and continues in free flight until impact occurs. The center of propulsion occurs at the CG of the complete legform, which is located near the damper.

The instrumentation required for the legform as proposed in the WG-17 report is a uniaxial tibia accelerometer located on the non-impacted side, a rotational potentiometer in the femur that measures the movement of the internal shear assembly, and a rotational potentiometer in the tibia that measures the movement of the femur with respect to the tibia.

Figure 3. Upper Legform.

The upper legform device is propelled using the same linear impactor that was just described for the complete legform test, but an adaptor plate is used to switch from a free-flight condition to a guided condition. The difficult design issue was that the propulsion unit, or piston, without the upper legform mass attached, could not weigh more than 3.65 kg. This is due to the WG-17 requirement that the impact mass and impact velocity be based upon the vehicle shape, which varies per vehicle. So the system parameters had to be flexible enough to cover both the free flight condition as well as the minimum and maximum guided conditions up to the heaviest propulsion mass and highest required impact velocity (11.65 kg at 40 km/h). This weight constraint effectively limited the acceleration stroke available to the free flight complete legform. In addition, concerns regarding the possibility of premature bending or deformation of the skin due to the level of acceleration required to reach the impact velocity resulted in a recommendation that the acceleration stroke be more than 1 m for the complete legform. However, the WG-17 report allows this stroke distance to be disregarded if the complete legform is supported at the inner leg through the skin up until release into free flight, which is why the locator pins were used.

The instrumentation of the upper legform consists of three strain gauges on the main beam to measure the strain at the lower, center, and upper portions of the form and two dynamic load cells that measure the force at the upper and lower portions of the form.

Figure 4. Child Headform (WG-17 version).

The child headform test uses a propulsion device that allows the form to impact the target point in a free-flight condition. The child headform is made up of an aluminum hemisphere with a diameter of 130 mm and has a 12.5 mm thick skin covering. Three uniaxial accelerometers (or one triaxial accelerometer) are mounted at the CG of the headform, where the propulsion also occurs.

Figure 5. Adult Headform (WG-17 version).

The adult headform uses the same impactor described for the child headform test. The adult headform is made of up an aluminum hemisphere with a diameter of 165 mm and has a 12.5 mm thick skin covering. Similar to the child headform, three uniaxial accelerometers are mounted at the CG of the headform, where the propulsion also occurs.

The differences between the WG-10 and the WG-17 impact devices include headforms made of a different material (aluminum instead of steel with a plastic front face) and the addition of the shear damper to the complete legform in order to eliminate the resonant frequency found in the undamped legform (WG-10). The upper legform did not have any changes between the two reports.

The initial interest in the US appears to be mostly in the complete legform and headform tests, mainly due to the difficulties encountered with the upper legform-to-bonnet leading edge test. Many groups are currently recommending that the upper legform-to-bonnet leading edge test be used solely for monitoring purposes, at least for the near future. This test also happens to be the most complicated of the tests to set-up. It involves variable masses, angles, and speeds that relate to the shape of the vehicle, which makes each test set-up unique. This is in contrast to the other WG-17 proposed test methods (complete legform, headforms) that include specific speed, angle, and mass impact parameters.

Calibration - Exact calibration corridors for the four impact forms have not yet been defined. WG-17 offers values in their report but it is emphasized that they are only preliminary values. The impact form manufacturers are still working on finalizing the acceptance corridors and test conditions for the various calibration tests.

To summarize the calibrations briefly, the adult and child headforms are calibrated using the same impact form. It consists of a flat aluminum face with a 70 mm diameter and a weight of 1.0 kg. Each headform is suspended from a cable and hit by the impact form at a velocity of 7 m/s (three hits per calibration). The criterion is that the peak resultant acceleration of each headform must fall within the range of 300 g's to 330 g's.

The upper legform is fired into a suspended tube as shown in Figure 3. The tube weighs 3 kg, with a diameter of 150 mm and 275 mm in length. The tube is positioned so that the upper legform impacts it at the intersection of the tube's longitudinal axis and the center of propulsion of the upper legform. The criteria include peak forces and bending moments and allowances for the peak differences between the upper and lower sensors. The complete legform has three calibration tests. Dynamically, it is suspended from a cable and hit by a hemispherical impact form with a diameter of 210 mm. Statically, the legform knee ligament batches and the potentiometers are tested together as a system in shear and bending scenarios using a cantilever-type metal tube to simulate the condition.

TEST PROCEDURE – Table 1 presents a summary of the WG-17 proposed tests and the criteria as well. One difference from the WG-10 report is that the WG-17 report added an optional test alternative for the complete legform-to-bumper test. This optional test is a horizontal, guided upper legform-to-bumper test and only applies to vehicles with a lower bumper height greater than 500 mm above the ground. This optional test is intended to more realistically simulate the struck pedestrian's lower extremity interaction with higher, raised vehicles such as sport utility vehicles (SUVs) or light trucks. This option uses the guided upper legform instead of the complete free flight legform because of the lack of tibia interaction seen when striking vehicles with higher bumpers.

Table 1. Summary of WG-17 Proposed Tests		
Test Name	Test Set-up	Criteria
Complete legform-to-bumper	40 km/h Mass=13.4 kg Imp. angle=0°	Max. dynamic knee bending angle of 15°; max. dynamic knee shearing displ. not to exceed 6 mm; tibia accel. less than 150 g's
Upper legform-to-bumper **(optional)**	40 km/h Mass=9.5 kg Imp. angle=0°	Sum of the impact forces not to exceed 5.0 kN; bending moment at any gauge shall not exceed 300 N-m
Upper legform-to-bonnet leading edge	Up to 40 km/h Variable mass Variable angle	Sum of the impact forces not to exceed 5.0 kN; bending moment at any gauge shall not exceed 300 N-m
Adult headform-to-bonnet top	40 km/h Mass=4.8 kg Imp. angle=65°	HPC not to exceed 1000
Child headform-to-bonnet top	40 km/h Mass=2.5 kg Imp. angle=50°	HPC not to exceed 1000

As indicated in Table 1, all of the tests are conducted at an impact speed of 40 km/h with the exception of the upper legform-to-bonnet leading edge test. In addition, the mass and impact angle for each test is specified, again with the exception of the upper legform-to-bonnet leading edge test where the impact mass and angle vary based upon the vehicle shape. Specifically, the bonnet leading edge height and the bumper lead measurement are two variables that affect the upper legform-to-bonnet leading edge test parameters.

The test criteria for each test are also noted in Table 1. The complete legform test has several test criteria including a maximum knee bending angle of 15°, maximum knee shearing displacement of 6 mm, and tibia acceleration below 150 g's. Both upper legform tests have the same test criteria. It is based upon the sum of the forces of the upper and lower load transducers as well as the bending moment of the strain gauges instrumented on the upper legform.

As far as the headform tests, the requirement is that the Head Performance Criterion, or HPC, must not exceed 1000. The HPC calculation is based upon the resultant acceleration time history and is the maximum of the equation, shown in Figure 6, calculated over time (with a maximum duration of 15 ms).

$$HPC = \left[\frac{1}{t_2 - t_1} \int_{t_1}^{t_2} a \, dt \right]^{2.5} (t_2 - t_1)$$

Figure 6. HPC calculation formula.

Details regarding the impact zones and reference lines for each type of test are presented in Figure 7.

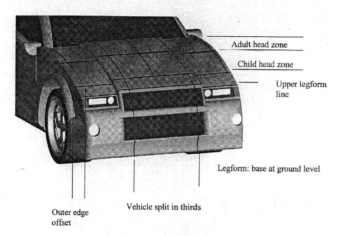

Figure 7. Impact locations and reference lines.

TEST DATA – Numerous evaluations have been conducted for all of the proposed tests detailed in the previous section using the WG-17 impact forms. Results of two of the tests, namely, the complete legform-to-bumper test and the adult headform-to-bonnet top test will be discussed. For each test, a typical passenger vehicle was used.

Complete Legform-to-Bumper Test – For this test, the impact speed, mass, and angle were adjusted to the specified levels previously described in Table 1. Knee shear displacement, knee angular displacement, and tibia acceleration were collected. Table 2 details the required data processing steps for this test.

Table 2. Summary of Data Processing Steps for Complete Legform-to-Bumper Test.	
Step	Process
1	Collect data set.
2	Define corridor of impact event.
3	Integrate acceleration to obtain velocity.
4	Scale the shear and bending curves to the required engineering units.
5	Filter shear, bending, and acceleration files to CFC 180.
6	Determine the maximum values for each file and determine the performance based upon the test criteria.

Figures 8-10 present the data collected for this test. The data indicates that the angular knee displacement exceeded the criteria limit. However, the peak acceleration of the tibia and the shear displacement of the knee were below the criteria.

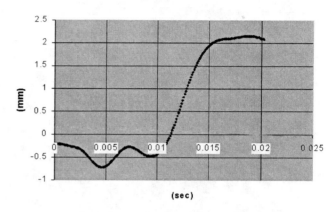

Figure 8. Shear displacement (knee).

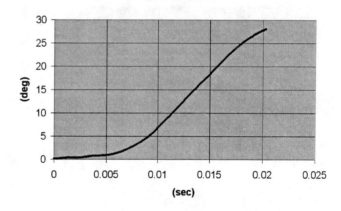

Figure 9. Angular displacement (knee).

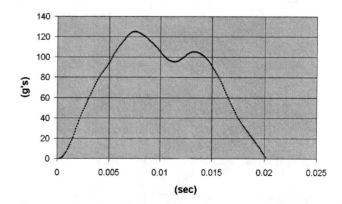

Figure 10. Tibia acceleration.

The data above has a CFC 180 filter applied to it based upon the recommendations of WG-17. The peak tibia acceleration was within the maximum limit of 150 g's as well as the shearing displacement (measured only 2 mm during the test). But, the knee angular displacement exceeded the limit of 15° by more than 10°.

Post-test photographs of the complete legform impact device are presented in Figures 11 and 12.

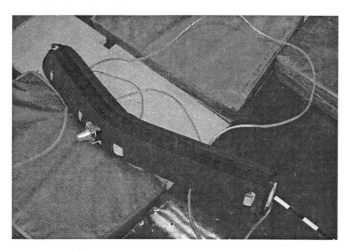

Figure 11. Post-test #1.

Figure 12. Post-test #2.

The photographs indicate that there was a large amount of angular deflection of the knee during the impact. This particular test maximized the amount of deflection allowed by the complete legform device, as seen in Figure 12. This indicates that the impact surface was relatively stiff only at the tibia or lower portion of the complete legform, and that the fibula or upper portion of the complete legform did not encounter any significant resistance until it impacted the hood area. This time delay between the vehicle contact with tibia and with the fibula contributed to the high angular displacement value obtained during this test.

Adult Headform-to-Bonnet Top Test - For this test, the impact speed, mass, and angle were adjusted to the specified levels (see Table 1). Laser positioning and

simulation software were used to ensure the first point contacted on the headform was at the intended target location. This was confirmed with the use of high-speed video. Headform triaxial acceleration data was collected. Figure 13 presents the resultant acceleration data plot and Figure 14 shows a post-test photograph of the impact target point.

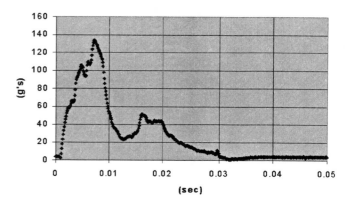

Figure 13. Adult Headform Acceleration Plot.

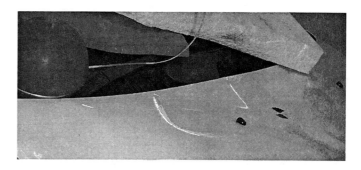

Figure 14. Post-test #1.

The peak of the head resultant acceleration, with a CFC 1000 filter applied, was 132 g's with an HPC of 723, which is below the limit of 1000. The photo shows that the deformation of the hood area surrounding the impact target location attributed greatly to the dissipation of the impact energy in this case. There were no rigid engine components underneath the hood at this particular impact area, which also allowed the deformation to occur.

CONCLUSION

As indicated earlier, until the impact forms and test criteria are finalized, the development of proper test procedures and equipment for Pedestrian Impact testing will continue. Specific calibration corridors and repeatable test methodologies are critical to the success of conducting these types of tests. Additional work may be necessary to continue to minimize variance in the sensor data. Issues such as the premature bending of the complete legform during propulsion and compensating for the change in trajectory of the headform due to gravity in free flight are quite important factors that affect the outcome of the test. Further studies will need to be undertaken to ensure that these issues are fully understood and are accounted for in both the test procedures and criteria.

ACKNOWLEDGMENTS

The authors of this paper would like to express appreciation to the associates of MGA and various customers who have contributed to the findings presented in this paper.

REFERENCES

1. European Experimental Vehicles Committee: Working Group 10 on Pedestrian Protection (1994). Proposals for methods to evaluate pedestrian protection for passenger cars: Final Report. EEVC, November 1994.
2. European Experimental Vehicles Committee: Working Group 17 Report on Improved Test Methods to Evaluate Pedestrian Protection Afforded by Passenger Cars (1998). EEVC, December 1998.
3. Code of Federal Regulations (CFR) Title 49 (Transportation) Chapter V (NHTSA, DOT) Part 571-201U, Upper Interior Head Impact Protection.
4. Code of Federal Regulations (CFR) Title 49 (Transportation) Chapter V (NHTSA, DOT) Part 571-203, Impact Protection for the Driver from the Steering Control System.

CONTACT

Helen A. Kaleto (Helen.Kaleto@mgaresearch.com)

Michael Worthington
(Michael.Worthington@mgaresearch.com)

Design Aspects of Energy Absorption in Car Pedestrian Impacts

Robert Kaeser and J.-Michel Devaud
Institute for Lightweight Structures
Swiss Federal Institute of Technology

ABSTRACT

The car-pedestrian impact is formulated as a design task for the engineer designing a car front. Pedestrian load tolerances are defined in an appropriate form for design. A two dimensional mathematical model with assumed deformable car surface is used to determine the movement of the pedestrian during the collision, the locations and velocities of impact. The dynamic load bearing capacity of materials and car body components is determined using an impact pendulum. With the described design procedure an experimental front was defined, built and tested in full scale crash tests. Results are presented. Emphasis is laid on simplicity of design methods as well in analysis as in experimental testing.

1. DESCRIPTION OF THE PROBLEM

Consider a car hitting a pedestrian. The position of the pedestrian, the velocity of the car and the first contact point on the car are given and may vary. The movement of the impacted pedestrian, further contact points on the car body and the forces which are acting between car body and pedestrian are the essential quantities related to injuries.

From investigation of real accidents the distribution of collision speeds and the contact regions on the car front are known. Fig 1 shows the cumulative frequency of collision speeds

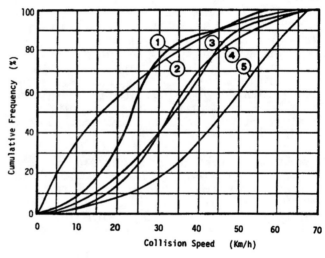

1. IAU, Zurich 1978, urban /10/
2. Danner, BRD / 7 /
3. Appel, BRD, urban / l /
4. Ashton, GB, urban / 3 /
5. Hutchinson,GB,urban,fatal /12/

Fig 1: Collision speed, results of different studies

from different studies. Fig 2 shows the frequency of contact points on the surface of current vehicles at collision speeds below and beyond 25 km/h. Most of the current production cars show a large variation of the form and of the local stiffness not only in longitudinal but also in transversal direction of the car front, which influences the movement of the pedestrian and the local forces, respectively, as well as the injuries.

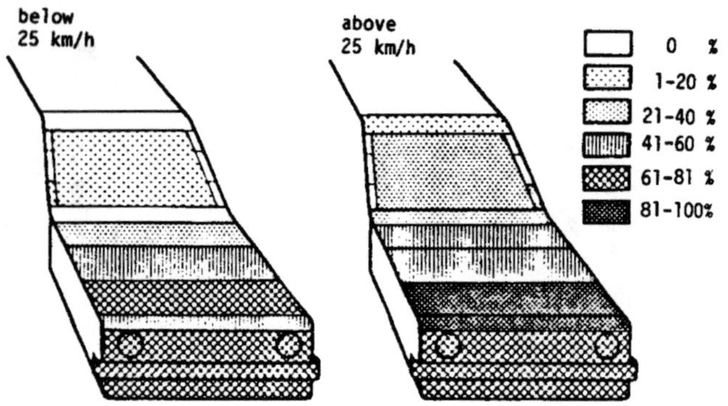
below
25 km/h
above
25 km/h
0 %
1-20 %
21-40 %
41-60 %
61-81 %
81-100%

Fig 2: Location of impact points on the car Front For impact speeds below and above 25 km/h /10/

We have no direct information on the contact Forces acting between the pedestrian and the car Front. In real accidents we observe injuries to the pedestrian and in Full scale tests accelerations of some parts of the dummy, as it is much easier to measure accelerations than contact Forces. Interpreting deformations on a car front or injuries to a pedestrian in order to determine contact Forces and time is not practicable.

The deformations on current production cars remaining after a collision with a pedestrian are in general very small. The force-deflection relationships For the contact points of the carbody are unknown and vary very much From point to point on the car surface.

To sum up we are able today to determine the movement of a given dummy during a collision with a given carfront, to determine accelerations at some points of the dummy, the points of contact and the permanent deformation of the car at those points. So Far we discussed the problem of an existing car hitting a pedestrian.

The engineer designing a new car Front, which will not hurt the pedestrian seriously at collisions up to a given velocity needs a different Formulation of the problem:

Given are a pedestrian with specified tolerance Levels of Forces and accelerations and some main dimensions of the car which shall be designed.

The shape and the stiffness characteristics of the car surface should be determined such that the defined pedestrian tolerance levels will not be exceeded during collision up to a chosen speed.

2. IMPACT ENERGY ABSORPTION

The relation between kinetic energy E, mass m, velocity v, deformation work W, Force F and deformation u

$$E = \frac{m \cdot v^2}{2} \qquad \text{and} \qquad W = \int F(u) \cdot du$$

cannot be applied without modification in the case of car-pedestrian impact as the contact zones and the masses involved are varying rapidly during the collision. The contact zones can be determined From experiments but it is difficult to judge how much mass is really accelerated at a specific contact point, since the inertia Forces of the adjacent body parts of the pedestrian enter into the equilibrium equations.

Experiments show the Following typical impacts of body parts with the car Front:

Impact of the bumper on the leg, impact of the leading edge of the bonnet to the Femur/pelvis region, impact of the head on or between bonnet and windscreen. At the moment of local impact the involved mass of the leg and the involved mass of the head can be estimated with sufficient accuracy. In the case of pontoon-form cars impact of the leading edge of the bonnet Follows in general directly after the impact of the bumper, so that the impact velocity of the femur/pelvis region is not influenced by car design. On the other hand the height of the impact zone of the leading edge of the bonnet has a large influence on the rotation of the body, the head impact velocity and location. Knowing impacted regions of the body and impact velocities from mathematical or experimental simulation, the effective masses can be estimated and the deformation energies of the different contact zones can be calculated. Stiffness of impacting area on the car-Front depends not only on static structural stiffness, but also on inertia Forces due to the involved mass of the impacting part of the carfront.

Assuming a force tolerance level, the force under which the impacting surface must deform is given. In the case of a stiff structure with a deformable foam layer on the impacting surface, the impacted parts of the body define the forms and extents of the deformed areas in the transversal direction. The stress level during compression and the extent of the impacted area in the longitudinal direction must be chosen such that impact loads will not exeed tolerable limits.

Materials and impacted components used in the design of energy absorbing carfronts must admit very large strain before failure. Steel metal sheets used generally for bonnets are very deformable at a low yield stress and show more than 30% strain at failure. Light metal alloys with low yield stress behave in a similar way. Plastics show very large strain at failure but very low yield stress so that their use for load bearing structural parts is limited. Fibre materials (glass, carbon, aramide) show very low strain at failure and are therefore inappropriate for energy absorption.

With plastics inexpensive foams can be made which have very interesting energy absorption characteristics but still present some technological problems (surface, production).

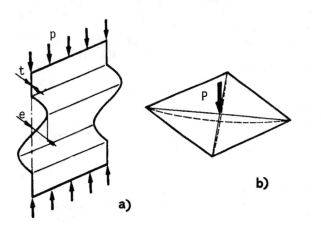

Fig 3: Use of metal sheet for energy absorption

With metal sheets, good energy absorption is only then realized, when applied forces are bending the sheet in the plastic range of the stress strain relationship (Fig 3).

In case a) the plastic moment m_p is

$$p \cdot e \quad \text{or} \quad \sigma_f \cdot \frac{t^2}{4}$$

so

$$p_{max} = \frac{m_p}{e} = \sigma_f \cdot \frac{t^2}{4 \cdot e}$$

For case b) there are no formulas as very little is known about the stiffness and deformation pattern of thin plates under large deformations in the plastic range. The resultant force from inertia and from bending (with or without additional stiffness from supports) should be in equilibrium with the inertia forces of the acceleated parts of the pedestrian according to the tolerance levels.

3. SYSTEMATIC DESIGN OF AN ENERGY ABSORBING CAR FRONT

In the first section the problem of the car-pedestrian impact was described from the point of view of the design engineer.

The aim of this study is to develop a procedure for the design of car fronts with acceptable behaviour in collisions with pedestrians. Important aspects of design have been discussed by different authors: Appel /2/, Ashton /4/, Brun /6/, Eppinger /8/, Harris /11/, Kühnel /15/, Pritz /18/ and Stcherbatcheff /20/.

The maximum velocity of impact which should not cause serious injuries must be defined. From statistical data it can be concluded that designing for a velocity up to 35 km/h will include the majority (84% in our investigation, Fig 1) of the urban collisions between cars and pedestrians. Furthermore at higher velocities the injuries should be less severe than with a design not considering pedestrian impacts.

The tolerance levels of the pedestrian must be specified in a form which is useful for the design engineer. A practicable way is to define the forces under which the car front surface must deform during impact with body parts, see Fig 4. The number of defined maximum impact forces should be small to

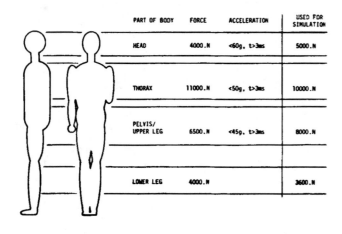

PART OF BODY	FORCE	ACCELERATION	USED FOR SIMULATION
HEAD	4000.N	<60g, t>3ms	5000.N
THORAX	11000.N	<50g, t>3ms	10000.N
PELVIS/ UPPER LEG	6500.N	<45g, t>3ms	8000.N
LOWER LEG	4000.N		3600.N

Fig 4: Load limits of pedestrian body parts

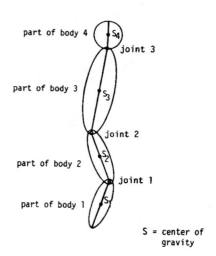

S = center of gravity

Fig 5: Model with 4 masses used for the mathematical simulation

limit the computing effort. The chosen values should represent the lower limits of the tolerance levels. A mathematical model is needed to determine the trajectory of the pedestrian and the locations of impact on the car surface. According to the required accuracy of the pedestrian's movement during impact different mathematical models have been used: Kramer /13/, Niederer /16,17/, Twigg /21/, and Wolff /22/. For design purposes it should be easy to use and lead with few iterations to a solution.

A twodimensional system of four or five masses has enough degrees of freedom to describe the essential phases of the collision. This means one mass for the lower legs, one for the upper legs, one or two for the trunk and one for the head (Fig 5). The influence of the arms which have a relatively low mass is neglected in this model. The masses are represented by elliptical segments connected by joints which are equipped with springs and damped to prevent increasing vibrations between the particular masses of the model. Furthermore angles of rotation can be limited and returning moments applied. It can be taken into account that the contact surfaces between body and car vary with the rotation of the body.

Some main geometrical data of the car front are given (height of bumper, leading edge of the bonnet, position of the windscreen, space occupied by motor and aggregates) or are chosen on the basis of previous experience with similar configurations. Given is the

initial state of the system. If all forces which act on the different parts are known, the accelerations of all parts can be calculated from the differential equations of motion. Integrating them over a time increment dt leads to velocities and with one more integration to the new coordinates at time t+dt.

The "contact" of the pedestrian with the car front is represented by impact loads which are defined as force-time functions acting on different points of the pedestrian model in defined directions. In that way, with given impact forces, the motion of the pedestrian model can be determined. The motion of the pedestrian can be improved varying the form of the car front within the limits of the given geometrical data. If the result is judged suitable, it is now the task of the designer to realize the defined force-deformation relationships with a real car front. This can be done using metal sheet, plastics, foam or structures of mixed materials.

4. TEST FACILITIES

Experimental testing is a necessary complement to the theoretical layout of deformable car components. In a first step the force-deflection relationships of appropriate materials and simple components used in the construction of the car body are determined under static and also under dynamic loads. In a second step complete body parts such as bumpers and bonnets are checked with regard to their behaviour under conditions similar to a car pedestrian collision. The impacting hammers used for that purpose must correspond in form, mass and impact velocity to the corresponding pedestrian body parts. These component tests enable to improve components before running full scale collision tests. Pedestrian impact test methodology has been discussed by Faerber /9/ and Eppinger /8/.

IMPACT PENDULUM - A pendulum is used to determine force-deflection relationships of energy absorbing elements like supported metal sheets or foam-specimens (Fig 6).

Fig 7: Impactor simulating lower leg hitting a foam made bumper

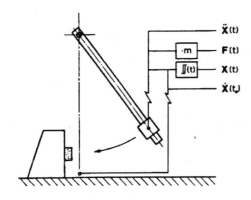

Fig 6: Pendulum for impact tests

By measuring the acceleration of the pendulum and integrating it twice, the dynamic force-deflection relation of a specimen during a defined impact (form, mass and initial velocity of the impact pendulum) can be obtained. Varying the mass and the drop height of the impacting pendulum, the impact energy can be varied over a wide range. The same test device is used to simulate the impact of the bumper as well as the leading edge of the bonnet (Fig 7). In this study, the width of the impacting, rounded pert of the pendulum is 160 mm for the upper part of the leg, and 100 mm for the lower part.

HEAD PENDULUM - With this test device the impact of the head on the bonnet is investigated (Fig 8). It is assumed that the head is moving like an unrestrained mass during the impact, neglecting coupling between chest and head.

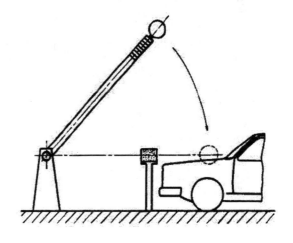

Fig 8: Test device for simulation of the head impact

An upright standing bar equipped with the head of a dummy on its upper end and with a hinge on the lower end falls down. Before the impact of the head takes place, the bar is stopped separately and the head strikes the bonnet. The neck is made of a series of light discs of plastic foam. During the fall head and discs are connected by a prestressed rubber cord. Directly before impact takes place the centrifugal force of the head exceeds the initial force in the rubber cord so that the foam discs dissociate and no longer transmit axial or shear forces.

Acceleration of the dummy head in the impact direction is measured with an accelerometer. The impact phase is filmed with a high speed camera to get information about deformations and vibrations of the bonnet. Impact velocity of the head of the test device used in this study was about 10 m/sec which is adequate to simulate a car-pedestrian collision speed of about 35 km/h. Deformation at the center of impact is measured with a dynamic displacement transducer.

The calculated movement and accelerations of the dummy can be checked by running tests on <u>full scale crash test facilities</u>.

5. RESULTS

MATERIAL AND COMPONENT TESTS - A large variety of <u>plastic foams</u> has been tested with the described impact pendulum. The aim of these tests was to find materials with good energy absorption at constant force level and without deformation remaining after the impact.

With regard to the defined conditions (loads, geometry) for the impacted parts of the car front, contact pressure during deformation on the leading edge of the bonnet and on the bumper is about 30 to 40 N/cm2 and about 25 N/cm2 in the head impact region on the bonnet and between bonnet and windscreen.

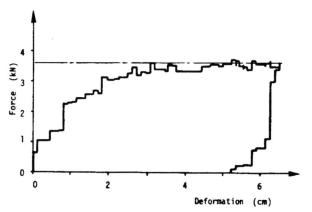

Fig 9: Dynamical force-deformation characteristic of a bumper of rigid PUR foam

Rigid foams made of polyurethane (PUR) and polyisocyanurate (PIR) show good energy absorption characteristics but deformations remain after impact (Fig 9).

Semiflexible foams made of PUR and PVC with densities from 50 to 120 kg/m3 fulfil the mentioned pressure requirement quite well. The use of materials with linear force deflection relationships for components results in large deflections and/or high maximal loads.

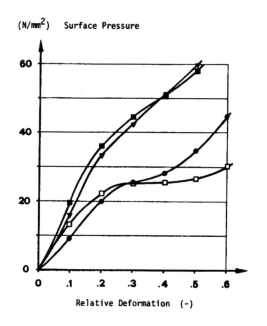

Fig 10: Deformation characteristics of PUR foams:
—□— rigid PUR foam

—●— semielastic PUR foam with density 120 kg/m3

—▼— increase of the deformation force at higher density (150 kg/m3)

—■— increase of the deformation force by varying chemical composition, density 120 kg/m3

With a systematic variation of the chemical composition of the foams, similar energy absorbing behaviour could be obtained with lower density (Fig 10). Currently produced semiflexible foams have no resistant skin so that they must be covered with a thin scratch-resistant coating. The influence of a polyurethane skin produced with the Reaction Injection Molding (RIM) technology on the force-deflection relationship of the foam core of a bumper is shown in Fig 11.

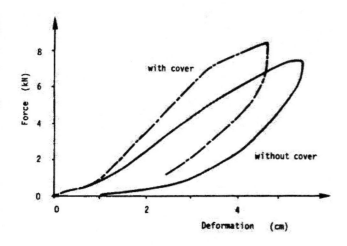

Fig 11: Energy absorption characteristic of a semiflexible PUR bumper with and without RIM-PUR cover

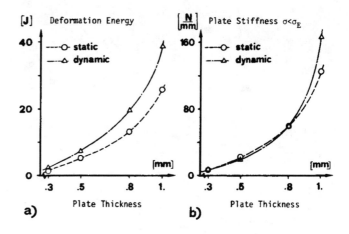

a)   b)

Fig 13: For thin metal sheets under static and dynamic loading with concentrated lateral Load at the center:
a) deformation energy
b) stiffness of the sheets in the elastic range

Current car fronts are mainly built with metal sheets. For better understanding of the energy absorption characteristic of that type of structure, a square metal sheet has been investigated under the action of static and dynamic concentrated loads in the middle of the thin plate with different boundary conditions, sheet thicknesses and impact energies.

clamped plates (but without in-plane forces on the boundaries, Fig 12). At impact velocities of 2 to 5 m/set the force deflecting the simply supported metal sheets was 40% higher during dynamic loading than under static loading. For metal sheets with clamped edges no significant difference could be found (Fig 13).

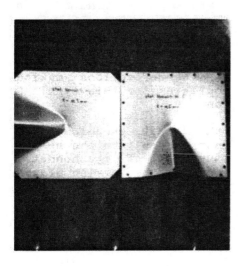

Fig 12: 2 metal sheets after static loading with concentrated lateral load at the center

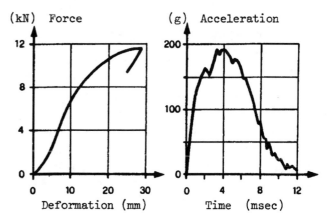

Fig 14: Characteristics of a simulated head impact with an impact velocity of 32 km/h on a Renault R4 bonnet (Fig 15)

Experiments have been carried out with simply supported as well as with

With different production cars <u>head impact tests</u> on the bonnet have been carried out with the head pendulum

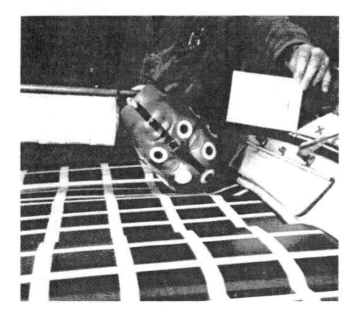

Fig 15: Position of the impacting headform and deformed bonnet after impact with 32 km/h

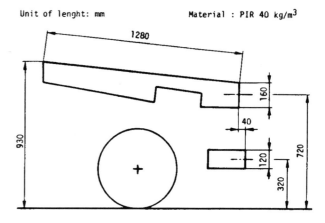

Unit of lenght: mm   Material : PIR 40 kg/m^3

Fig 16: Geometry of the energy absorbing parts of the car front used for mathematical and experimental simulation

(Fig 15). The measured accelerations of the head are on average very high, on the one hand because of the excessive stiffness of the bonnet itself, on the other hand because of the lack of space under the hood which limits deformations (Fig 14).

EXPERIMENTAL CHECK OF THE DESIGN PROCEDURE - The described design procedure has been applied to build an experimental car front which then has been tested on the crash test facility of the "Arbeitsgruppe für Unfallmechanik" at the Universities of Zurich. The following parameters have been chosen for the mathematical simulation:

- masses, geometry and defined tolerance levels of the 50 percent male pedestrian standing with his side facing the vehicle front
- impact velocities: 25 and 35 km/h.

The assumed tolerance loads used for this simulation follow from Fig 4.

The shape of the experimental front was determined using the mentioned simulation program, such that head impact occurred in a way which seemed the best tolerable on a compact front. Simulation of the pedestrian kinematics gave as a result the necessary deformations at the collision contact points on the car surface. Based on obtained values an experimental front has been built (Fig 16). To realize the required force-deflection relationships, a rigid PIR foam with nearly constant force level during deformation was used. The brittle failure under compression of this material allows an exact identification of the collision contact points on the car front surface and a good estimate of the deflections making it very appropriate for this type of investigation.

With the impact pendulum previous tests of components like the bumper and the leading edge of the bonnet have been carried out by varying dimensions and materials until the required force-deflection relationship was obtained. The amount of impact energy has been chosen equal to the work done by impact forces on the car front in the mathematical simulation. With the finally resulting experimental front eleven full scale tests have been run with a modified PART 572 Dummy at collision speeds of 25 and 35 km/h.

To prevent the movement of the left arm influencing the head impact, it was fixed on the dummy's back. The dummy was first impacted on his left side under an angle of 20 degrees towards or sway from the vehicle front. In that way on impact was obtained either on the chest or on the back.

During the tests the following quantities were measured:

- accelerations of head, chest, pelvis and knee of the dummy (all with triaxial accelerometers)
- movement of the dummy with high speed cameras and therefrom the trajectory and the velocity-time curve of the head

Fig 18: Deformation of the leading edge of the bonnet after impact at 25 km/h

Fig 17: Experimental front with marks of an impact at a collision speed of 35 km/h

Additional measurements in some experiments were:

- the force acting on the leading edge of the bonnet in the driving direction of the car
- deformation of the leading edge of the bonnet during impact.

After each test deformations remaining on the vehicle front were measured (Fig 17,18).

COMPARISON OF THE MATHEMATICAL SIMULATION WITH THE EXPERIMENT - On the one hand the calculated and measured movements of the dummy and on the other hand the impact characteristics are compared.

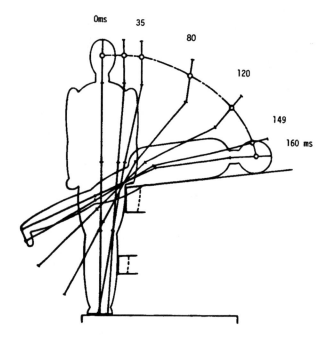

Fig 19: Kinematics of the used
pedestrian model, math.
simulation at 25 km/h

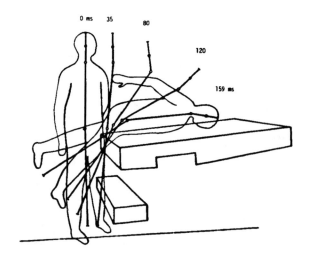

Fig 20: Overall motion of dummy
PART 572 during test
at 25 km/h

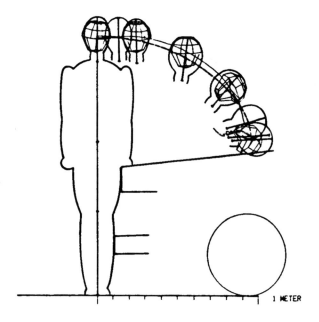

Fig 21: Head trajectories at
25 km/h:
----from simulation
from experiment (with
-methods from /17/)

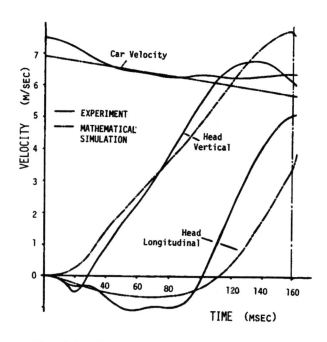

Fig 22: Horizontal and vertical speed
components of the head motion
from math. simulation and
from experiment at 25 km/h

At impact velocities of 25 km/h the movement of the dummy corresponds very well with the mathematical simulation (Fig 21,22). Point of time, location and velocity of the head impact have been reproduced well. In the experiments run at an impact velocity of 35 km/h there are two variants of the last phase of the dummy's movement prior to head impact depending on the direction in which the dummy is rotating (Fig 27,28). The dummy which is facing the impacting front under 20 degrees is falling continuously onto the bonnet until the head is impacting. The dummy which is impacted under 20 degrees from behind is leaning on his right arm on the bonnet. Thus the upper part of the body is retarded, the head rotates causing the neck to bend. The head hits the bonnet under a large angle and strongly reduced velocity (Fig 26).

Even under these special conditions the general sequence of motion is very similar to the calculated one. With both configurations the component of velocity of the upper part of the body relative to the car front is smaller in the experimental tests than in the mathematical simulation (in the longitudinal direction of the car in the moment of chest and head impact).

Consequences of impact as acceleration curves of the dummy body parts and deformations on the car front have been compared to judge the validity of the design procedure concerning the aim to stay within determined load limits with the corresponding acceleration values (Fig 23,24). In the cases of leg impact on the bumper and head impact on the bonnet the deformations of the front and acceleration curves compare well.

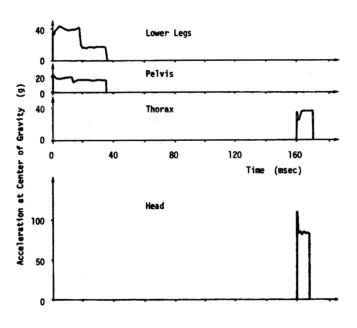

Fig 23: Accelerations of body parts, math. simulation at 25 km/h

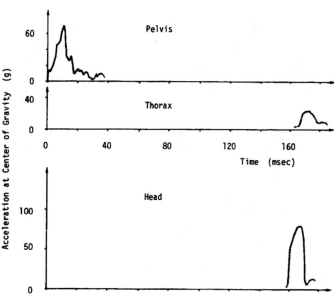

Fig 24: Accelerations of body parts during experiments at 25 km/h

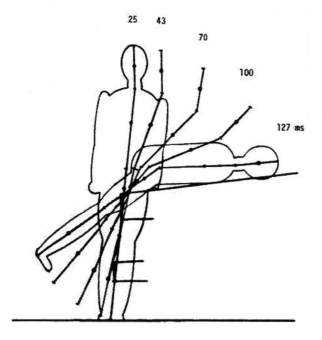

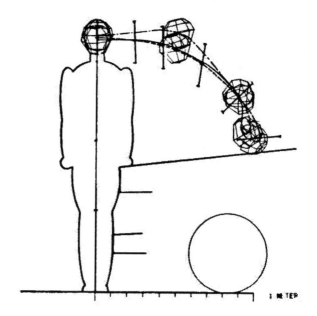

Fig 25: Kinematics of pedestrian modell. Math. simulation of an impact at 35 km/h

Fig 26: Head trajectories at 35 km/h from:
--- math. simulation
—— experiment, rotation on the back
—·— experiment, rotation on the chest

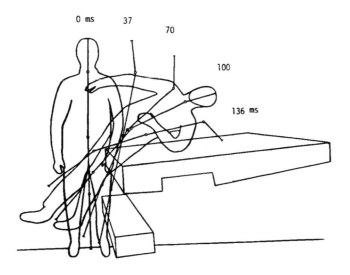

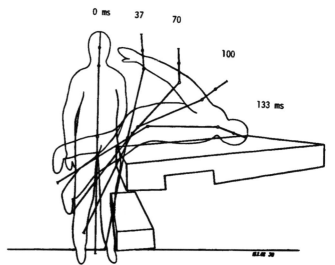

Fig 27: Dummy motion with rotation on the back. Note interference of the elbow. Experiment at 35 km/h

Fig 28: Dummy motion with rotation on the chest. Experiment at 35 km/h

Impact of the femur/pelvis region with the leading edge of the bonnet results in a deformation which in teats is only one half the expected value, while the curve of the measured acceleration at the pelvis is very uneven and has a mean value which is twice the expected one. Deformation pattern of the leading edge of the bonnet, especially the width of the defored zone is the same in dummy tests as in component tests so that it can be supposed that the impact force was of the right order of magnitude. It can therefore be concluded that, during the experimental impact of the leading edge of the bonnet, the effective accelerated mass of the pedestrian was only half the assumed mass in the mathematical simulation. It seems that by introducing one more hinge in the pedestrian model the flexibility of the system pelvis/vertebral column would be simulated in a better manner. Apart from this detail which can easily be modified the accuracy of the results of the mathematical simulation is very satisfactory for design purposes.

6. CONCLUSIONS

From a two dimensional mathematical model a prediction of the pedestrian's motion and of the time, location and velocity of head impact can be made which is good enough for design purposes.

A car front designed with regard to collision with pedestrians will show larger deformations under impact than a current car. The size of the deformations depends on the defined pedestrian tolerance levels and the assumed collision velocity.

Standardisation would be useful for:
- maximum impact velocity which should be considered in design
- tolerable loads and accelerations of the pedestrian which should be considered in design.

Impact test devices for leg and pelvis/femur impact should be improved before standardisation.

Energy absorption of thin metal sheets in plastic bending should be investigated thoroughly as this are often used components in car body construction.

ACKNOWLEDGMENTS

The research work discribed in this paper was supported in part by the Swiss Foundation for the Prevention of Traffic Accidents.

8. REFERENCES

/1/ Appel H., Stürtz G., Gotzen L.: Influence of Impact Speed and Vehicle Parameters on Injuries of Children and Adults in Pedestrian Accidents. Proceedings IRCOBI Conference (9.9. - 11.9.1975)

/2/ Appel H., Kühnel A., Stürtz G., Glöckner H.: Pedestrian Safety Vehicle - Design Elements - Results of In-Depth Accident Analyses and Simulation. Proceedings 22nd AAAM - Conference, 132 - 153 (1978)

/3/ Ashton S.J., Pedder J., Mackay G.: A Rewiew of Riders and Pedestrians in Traffic Collisions, Proceedings of IRCOBI Conference (7.9. - 9.9.1977)

/4/ Ashton S.J., Mackay G.M.: Car Design for Pedestrian Injury Minimization, 7th International Technical Conferenz on Experimental Safety Vehicles, Paris, June 6-9, (1979)

/5/ BASt: Biomechanische Belastungsgrenzen. Unfall- und Sicherheitsforschung im Strassenverkehr 3, Köln (1976)

/6/ Brun F., et al: A Sythesis of Available Data for Improvement of Pedestrian Protection, 7th International Technical Conference on Experimental Safety Vehicles, Paris, June 6-9, (1979)

/7/ Danner M., Langwieder K.: Collision Characteristics and Injuries to Pedestrians in Real Accidents. Proceedings 7th International Technical Conference on Experimental Safety Vehicules, London (1979)

/8/ Eppinger R.H., Pritz H.B.: Development of a Simplified Vehicle Performance Requirement for Pedestrian Injury Mitigation, 7th International Technical Conference on Experimental Safety Vehicles, Paris, June 6-9, (1979)

/9/ Faerber E., Glaeser K.-P.: Considerations on a Standardised Pedestrian Test Methodology. Proceedings of IRCOBI Conference, Köln (1982)

/10/ Gaegauf M., Niggli E., Wehren A.,: Fussgängerunfallgeschehen in der Stadt Zürich, Interdisziplinä-re Arbeitsgruppe für Unfallmechanik, Univerität Zürich (1981)

/11/ Harris J., Radley C.P.: Safer Cars for Pedestrian, 7th International Technical Conference on Experimental Safety Vehicles, Paris, June 6-9, (1979)

/12/ Hutchinson T.: Factors Affecting the Times Till Death of Pedestrians Killed in Road Accidents. Injury 6, 208 - 212 (1975)

/13/ Kramer M.: Ein einfaches Modell zur Simulation des Fahrzeugfussgängerunfalles. ATZ 76 Stuttgart, BRD (1974)

/14/ Kramer M.: Bestimmung der äquivalenten Verletzungsschwere in experimentellen Fussgänger-Fahrzeug-Unfällen. Fortschrittsbericht VDI 17 Nr 3 (1977)

/15/ Kühnel A. et al: First Step to a Pedestrian Safety Car. 22th STAPP Car Crash Conference (1978)

/16/ Niederer P: Computerized Simulation and Reconstruction of Car-Pedestrian Accidents. Int. Microf.J.leg.Med. 11 No2 (1976)

/17/ Niederer P., Schlumpf M., Hartmann P.A.: The Reliabilitzy of Antropometric Test Devices, Cadavers and Mathematica1 Models as Pedestrian Surrogates, SAE Congress, Detroit, MI, (Febr 83)

/18/ Pritz H.B.: Vehicle Design for Pedestrian Protection, 7th International Technical Conference on Experimental Safety Vehicles, Paris, June 6-9, (1979)

/19/ Stcherbatcheff G. et al: Dissipation d'énergie par corps creux en tôle mince. Mécanique, Matériaux, Electricité 341/342 (1978)

/20/ Stcherbatcheff G. et al: Pedestrian Protection-Special Features of the Renault EPURE, 7th International Technical Conference on Experimental Safety Vehicles, Paris, June 6-9, (1979)

/21/ Twigg D,W. et al: Optimal Design of Automobiles for Pedestrian Protection. SAE Paper No 770094 (1977)

/22/ Wolff C.: Experimentelle und rechnerische Simulation von Fussgängerunfällen. TU Braunschweig (1976)

The Role of the Vehicle Front End in Pedestrian Impact Protection

Samuel Daniel, Jr.
National Highway Traffic Safety Administration

THE NATIONAL HIGHWAY TRAFFIC SAFETY ADMINISTRATION (NHTSA) has been conducting research activities aimed at reducing the number of pedestrian and cyclist fatalities and the severity of injuries sustained by these highway users resulting from accidents. There are two basic approaches presently being used by NHTSA to achieve this objective, one is through reducing the number of accidents (crash avoidance) and the other involves modifying vehicles that collide with pedestrians and cyclists to mitigate injury consequences (crashworthiness).

The term "cyclist" includes operators and passengers of all vehicles with three wheels or less, motorized and non-motorized. Cyclists involved in collisions with motor vehicles experience kinematics and injury patterns similar to pedestrians and will thereby benefit from pedestrian protection oriented vehicle modifications (12).*

---

* Numbers in parentheses designate References at the end of the paper.

PROBLEM IDENTIFICATION

The Pedestrian Injury Causation Study (PICS) has been used along with the Fatal Accident Reporting System (FARS) and the National Accident Sampling System (NASS) to generate statistics for pedestrian accidents. These files have been combined to project national accident and injury statistics.

There are approximately 137,000 annual accidental collisions between pedestrians and motor vehicles resulting in 8,000 fatalities (11). There are also about 190,000 motorcyclists involved in reported accidents annually with about 4,700 fatalities (12). Table 1 gives a breakdown of all motor vehicle accident related fatalities from the FARS data file which shows that 27% of the total annual fatalities are pedestrians and cyclists. Approximately 84% of the vehicles involved in pedestrian collisions are passenger cars according to PICS. Pedestrian collisions with the front surface structure of passenger cars, which includes the bumper, grille and hood edge, hood and fenders, produce 52% of the

---

ABSTRACT

National Annual Accident Statistics for pedestrians and cyclists are presented along with a breakout of injuries matched with vehicle impact location and other causation factors. Results of full-scale accident simulation tests using surrogates to investigate the relationship between impact response and injury severity for several body regions is presented. Accident investigation data, full-scale accident simulation research test data, and computer modeling data aimed at determining the effects of vehicle front surface structure geometry on pedestrian injury patterns and severity are presented. The general approach to pedestrian injury mitigation through vehicle modification is discussed. A discussion of motorcyclist injury patterns is also presented.

TABLE 1

1979 FATALITY DISTRIBUTIONS (FARS)

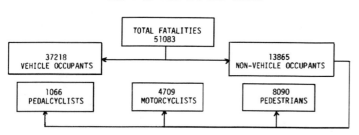

societal cost or harm of pedestrian accidents as can be seen in Table 3. Body regions are matched with injury causation contacts in Table 2 and the combinations are presented as a percentage of the total societal cost or harm associated with pedestrian accidents.

It can also be seen in Table 2 that injuries to the head, chest, abdomen and legs contribute significantly to the societal cost of pedestrian accidents and that adult (greater than 10 years of age) accident victims receive

Table 2 (From: Ref. 21)
Percent Distribution of Total Harm to Pedestrians

Vehicle Contact With Adults	PEDESTRIAN BODY REGION							
	ALL	HEAD	CHEST	ABDOMEN	NECK	PELVIS	UPPER X	LOWER X
1 Bumper	9.36	0.07	0.13	0.01	0.01	0.01	0.00	9.15
2 Grille & H. Edge	14.34	0.01	5.78	7.00	0.00	0.85	0.23	0.47
3 Hood & Fenders	20.96	7.56	6.16	3.62	1.48	0.36	0.94	0.83
4 Windshield	3.98	3.63	0.20	0.00	0.01	0.00	0.14	0.00
5 Side	0.98	0.36	0.11	0.24	0.00	0.04	0.13	0.09
6 Under	8.58	2.17	3.69	2.24	0.02	0.01	0.11	0.33
7 Other & Rear	2.41	0.01	0.22	1.45	0.00	0.22	0.04	0.45
8 Accessories	0.31	0.19	0.03	0.00	0.00	0.01	0.08	0.01
9 Noncontact	7.47	0.00	0.62	0.34	3.96	0.01	1.11	1.42
10 Road	10.91	7.79	0.33	0.45	0.25	0.11	1.37	0.61
All Adults	79.30	21.80	17.27	15.35	5.73	1.62	4.17	13.36
With Children (10 Yrs Old Or Less								
11 Bumper	2.66	0.01	0.07	1.10	0.00	0.09	0.04	1.34
12 Grille & H. Edge	3.92	1.25	0.79	1.48	0.21	0.04	0.14	0.02
13 Hood & Fenders	2.46	1.55	0.53	0.01	0.21	0.04	0.09	0.04
14 Windshield	0.00	0.00	0.00	0.00	0.00	0.00	0.00	0.00
15 Side	0.59	0.53	0.00	0.00	0.00	0.00	0.01	0.05
16 Under	2.42	0.62	0.37	0.37	0.13	0.00	0.04	0.89
17 Other & Rear	2.07	0.17	0.05	1.21	0.00	0.12	0.38	0.15
18 Accessories	0.10	0.01	0.04	0.00	0.00	0.00	0.04	0.01
19 Noncontact	2.18	0.00	0.00	0.44	1.26	0.00	0.07	0.41
20 Road	4.29	3.09	0.06	0.07	0.21	0.05	0.45	0.36
All Children	20.70	7.22	1.92	4.64	2.02	0.35	1.25	3.27
With All Pedestrians								
21 Bumper	12.02	0.07	0.20	1.11	0.01	0.10	0.04	10.50
22 Grille & H. Edge	18.27	1.26	6.57	8.48	0.21	0.89	0.37	0.49
23 Hood & Fenders	23.41	9.11	6.69	3.63	1.69	0.40	1.03	0.87
24 Windshield	3.98	3.64	0.20	0.00	0.01	0.00	0.14	0.00
25 Side	1.57	0.88	0.11	0.25	0.00	0.04	0.14	0.14
26 Under	11.00	2.78	4.06	2.61	0.16	0.02	0.15	1.22
27 Other & Rear	4.48	0.19	0.27	2.65	0.00	0.34	0.42	0.60
28 Accessories	0.41	0.20	0.07	0.00	0.00	0.01	0.12	0.02
29 Noncontact	9.65	0.00	0.62	0.78	5.22	0.01	1.18	1.83
30 Road	15.21	10.87	0.40	0.51	0.46	0.17	1.82	0.97
All Pedestrians	100.00	29.02	19.19	20.02	7.75	1.97	5.41	16.64

about 70% of the harm although statistics indicate that 40-50% of the accident victims are children. Table 4 shows that pedestrian harm is particularly sensitive to impact speed in the 25-30 mph range with 28% of the harm occurring in this speed range and 76% of the harm resulting from collisions with impact speeds of 30 mph or less. Table 5 gives hypothetical reductions in harm assuming a AIS unit reduction in injury severity for all injuries resulting from contact with the various vehicle components.

The data in Tables 2-5 indicate that the pedestrian injury problem is diverse with about 50% of the societal cost attributable to pedestrian contact with the roadway and areas of the vehicle that cannot be practically modified. The data in Table 5 also points out the fact that severity reductions of 1 AIS unit for injuries resulting from frontal structure contacts may achieve a significant reduction in the harm of pedestrian accidents and pedestrian fatalities, 38% and 26%, respectively.

The National Electronic Injury Surveillance System (NEISS) has been used along with NASS and a special study done by the Univ. of Southern California (USC) to develop statistics for motorcyclist injuries (12). When all motorcyclist injuries are considered, head and lower leg injuries are the most prevalent in terms of injuries per accident as can be seen in Table 6. For motorcycle crashworthiness evaluation, only injuries attributable at least in part to interaction with the cycle are considered for body regions other than the head. The breakdown of injuries by body region as a percentage of total harm are given in Table 7. The USC file was used for this evaluation.

APPROACH

The vehicle oriented approach to the pedestrian and cyclist injury problem involves utilizing data from multi-disciplinary accident studies to quantify the various injury types by

Table 3 (From: Ref. 21)
Resolution of Harm to Pedestrians into the Injury Severity Spectrum

	Car Component	Body Region	Adult or Child	Harm % of Total	Harm Distribution, Percent As a Function of Injury Severity		
					AIS=1	AIS=2 to 5	AIS=6
1	Bumper	Lower Extremities	Adult	9.15	5.5	94.5	0.0
2	Hood & Fenders	Head	Adult	7.56	1.9	87.2	10.9
3	Grille & Hood Edge	Abdomen	Adult	7.00	0.0	96.6	3.4
4	Hood & Fenders	Chest	Adult	6.16	0.1	89.3	10.6
5	Grille & Hood Edge	Chest	Adult	5.78	0.3	96.1	3.6
6	Hood & Fenders	Abdomen	Adult	3.62	0.4	99.6	0.0
7	Hood & Fenders	Head	Child	1.55	2.8	77.0	20.2
8	Grille & Hood Edge	Abdomen	Child	1.48	2.0	98.0	0.0
9	Hood & Fenders	Neck	Adult	1.48	0.0	2.5	97.5
10	Bumper	Lower Extr.	Child	1.34	14.9	85.1	0.0
11	Grille & Hood Edge	Head	Child	1.25	3.6	96.4	0.0
12	Bumper	Abdomen	Child	1.10	0.7	99.3	0.0
13	Hood & Fenders	Upper Extr.	Adult	0.94	21.3	78.7	0.0
14	Grille & Hood Edge	Pelvis	Adult	0.85	14.8	85.2	0.0
15	Hood & Fenders	Lower Extr.	Adult	0.83	13.4	86.6	0.0
16	Grille & Hood Edge	Chest	Child	0.79	3.7	96.3	0.0
17	Hood & Fenders	Chest	Child	0.53	2.8	97.2	0.0
18	Grille & Hood Edge	Lower Extr.	Adult	0.47	30.2	69.8	0.0
	Subtotal			51.88	4.1	88.6	7.3
19	Windshield	All	Both	3.98	3.4	86.2	10.4
20	Side	All	Both	1.57	9.4	90.6	0.0
21	Under	All	Both	11.00	0.9	73.0	26.1
22	Other & Rear	All	Both	4.48	6.1	84.6	9.3
23	Accessories	All	Both	0.41	28.6	71.4	0.0
	All the Above			73.32	4.0	85.8	10.2
	All Other Contacts with Vehicle		Both	1.82	4.1	88.6	7.3
	All Non-Contact		Both	9.65	0.3	50.1	49.6
	All Road Contacts		Both	15.21	21.5	69.0	9.5
	ALL			100.00	6.6	79.9	13.5

Table 4 (from: Ref. 21)
Resolution of Pedestrian Harm into the Impact Severity Spectrum

	Contact	Body Region	Adult/Child Split/Percent	Harm % of Total	Cumulative Distribution of Harm (%) Over Impact Severities. mph					
					15	20	25	30	40	50
1	Bumper	Lower Extremities	87/13	10.50	39	57	66	82	90	98
2	Hood & Fenders	Head	87/17	9.12	8	14	33	82	93	100
3	Grille & Hood Edge	Abdomen	83/17	8.48	17	25	41	73	88	98
4	Hood & Fenders	Chest	83/8	6.69	8	13	19	58	81	94
5	Grille & Hood Edge	Chest	88/12	6.57	21	26	33	97	99	100
6	Hood & Fenders	Abdomen	100/0	3.63	1	4	15	48	70	78
7	Hood & Fenders	Neck	88/12	1.68	2	2	2	88	88	100
8	Grille & Hood Edge	Head	1/99	1.26	7	38	95	100	100	100
9	Bumper	Abdomen	1/99	1.11	50	52	66	71	100	100
10	Hood & Fenders	Upper Extremities	91/9	1.02	42	55	64	82	86	100
11	Grille & Hood Edge	Pelvis	95/5	0.90	42	53	63	87	90	99
12	Hood & Fenders	Lower Extremities	96/4	0.87	34	37	44	46	93	100
13	Grille & Hood Edge	Lower Extremities	95/5	0.49	48	65	66	84	85	100
	Subtotal		83/17	52.32	20	29	41	77	89	97
14	Windshield	All	100/0	3.98	6	24	35	51	73	90
15	Side	All	62/38	1.57	31	39	55	64	87	96
16	Under	All	78/22	11.00	34	44	49	72	88	100
17	Other & Rear	All	54/46	4.48	38	48	58	77	88	95
18	Accessories	All	75/25	0.42	32	49	57	61	99	99
	All the Above		81/19	73.77	22	33	43	74	88	96
	All Other Contacts with Vehicle		83/17	1.38	20	29	41	77	89	97
	All Non-Contact		77/23	9.65	17	33	43	69	87	99
	All Road Contacts		72/28	15.20	47	63	76	85	94	99
	ALL		79/21	100.00	26	37	48	76	89	97

Table 5 (From: Ref. 21)
Pedestrian Harm Reduction and Fatality Reduction Effectiveness Obtained from an Hypothetical Reduction of Injury Severity by One AIS Unit for All Injuries Cauused by shown Contacts at Impact Severities Below Shown Values

Injuring Contact	Impact Speed, mph	Harm % of Total	Fatality % of Total
Bumper	20 mph	5.81%	3.46%
	25 mph	6.72	5.03
	30 mph	8.31	7.55
	All	10.44	9.42
Grille and Hood Edge	20 mph	4.16	4.51
	25 mph	6.00	6.13
	30 mph	11.10	6.86
	All	12.94	7.33
Hood and Fenders	20 mph	2.64	2.35
	25 mph	4.32	4.16
	30 mph	10.22	7.32
	All	14.82	9.68
All Vehicle Contacts	20 mph	21.10	19.71
	25 mph	27.26	27.05
	30 mph	44.19	37.83
	All	57.76	48.01
All Sources	20 mph	28.51	23.98
	25 mph	36.16	32.33
	30 mph	53.90	44.38
	All	68.80 Reduction	56.55 Reduction

102

Table 6 (From: Ref. 12)
Percent Insured Cyclists/Injured Body Region

	Head	Chest	Abdomen	Pelvis/Upper Leg	Lower Leg	Sample Size
NASS	40	38	6	14	75	150, 292
USC	64	31	8	17	96	1,027

frequency and severity per accident and identifying the vehicle contact points associated with each injury. Basic biomechanical research is then conducted to investigate the relationship between the independent collision parameters, such as impact speed and vehicle geometry and the resultant injury severity. Full-scale accident simulation using cadavers for pedestrian surrogates is the primary method of obtaining this type of information.

Anthropomorphic dummies and less complex devices designed to simulate the impact response of a body region or segment are also used as pedestrian surrogates in accident simulation research tests, primarily to study pedestrian kinematics and evaluate the relative performance of various vehicle designs. Full scale dummies and body segment simulators can be used to predict injury severity if they can be calibrated with human injury response data through accident reconstruction or reconstruction of impact tests using cadaver surrogates.

When impact injury levels for production vehicles are sufficiently established and impact response parameters which can be used as injury severity indicators are identified, an iterative process can be initiated aimed at developing vehicle components that significantly reduce the injury severity levels of the impacted body regions. All three types accident victim surrogates are used in accident

simulation tests during this process, along with analytical tools such as computer based math simulations. Modifications exhibiting promising impact response characteristics are then further investigated for mass production feasibility. Compliance criteria, which are based on the performance of the modified vehicles, are then established and compliance procedures developed.

BIOMECHANICAL RESEARCH

Injury statistics (Table 2) indicate that direct contact with the front surface structure of passenger cars is the source of a high percentage of the overall pedestrian injuries and the consequences of motor vehicle collisions with pedestrians. Researchers, primarily in the U.S. and Europe, have been conducting accident simulation tests over the past eight years to quantify the front structure in terms of injury response and to investigate injury mitigating structural modifications. Impact response data from accident simulation research tests using production vehicles and vehicles equipped with pedestrian oriented modifications is highlighted in this section.

Several pedestrian oriented front structure modifications have been fabricated and tested. They include softer bumpers using honeycomb metal or urethane foam for the facebar. The grill/hood edge area has been modified, primarily by eliminating sharp contours at the hood edge and vertical front corners and through the use of soft materials. The area at the base of the windshield (cowl), the windshield frame and engine compartment components that are close to the hood have also been modified to reduce pedestrian injury severity.

Table 7 (From: Ref. 21)
Motorcyclists Harm

Motorcyclists Body Region	Percent of Total Harm	Percent of Total Harm Reduction Obtained by Delta AIS of One at Crash Speeds Under (mph)				Percent of Total Harm Reduction Obtained by Eliminating All Injury at Crash Speeds Under (mph)			
		20	30	40	All	20	30	40	All
Head	35.7	2.0	6.8	14.8	20.7	4.4	11.0	25.4	35.7
Knee & Lower Leg*	5.9	0.9	1.9	3.5	4.5	1.0	2.4	4.7	5.9
Upper Leg*	0.6	0.0	0.1	0.5	0.6	0.0	0.2	0.5	0.6
All Other	57.8	-	-	-	-	-	-	-	57.8
ALL	100.0	-	-	-	-	-	-	-	100.0

*Requires at least one contact with the motorcycle, i.e., excludes injuries incurred exclusively by contact with objects other than the motorcycle.

PEDESTRIAN-BUMPER INTERACTION - The bumper surface location of passenger cars produced for sale in the U.S. since 1973 is fairly consistent in location, geometry, and mechanical properties. The bumper facebar is the forwardmost exterior surface component on passenger cars, extending the entire width of the vehicle with the vertical centerline at 17-19 inches above the ground and a vertical height of 4-7 inches. The facebar is typically a single homogenous unit of steel or aluminum, although high density polyurethane foam has been used in recent years on some models. The use of thin polyurethane fascia coverings for steel and urethane facebars has been increasing over the last 3 years, particularly on domestic models.

Approximately 12% of the societal cost of pedestrian accidents is attributed to the consequences of injuries caused by direct contact between the pedestrian and the front bumper. Researchers in the field of pedestrian injury mitigation have conducted numerous research programs to establish baseline vehicle impact injury severities, and to identify impact response parameters which indicate or predict injury severity.

The Battelle Columbus Laboratories (BCL) of Columbus, Ohio, conducted 25 full-scale accident simulation research tests using unembalmed cadavers as accident victim surrogates (1)(2). The lower leg injury produced by bumper impact was observed and several impact response parameters of the lower leg were recorded or calculated. The maximum acceleration of the knee was determined to be an important indicator of lower leg injury severity. The magnitude of the peak knee acceleration response for baseline vehicle impacts at speeds of 10-30 mph was roughly between 75 and 500 times greater than the acceleration of gravity (G's). The lower leg injury severity observed was between AIS 0 and 4.

Various types of modified bumpers were also used in similar tests and it was determined, based on these tests (Figure 1), that lower leg injury severity begins to change from AIS 2 to AIS 3 or greater for impact responses in excess of 100 G's. Figure 2 shows knee impact acceleration responses for dummy surrogates using production and modified bumper systems in full-scale impact tests. The data indicates that acceleration levels can be limited to 100 G's or less for impact speeds up to 25 mph with appropriate bumper modifications.

The bumper strikes the lower leg of adult

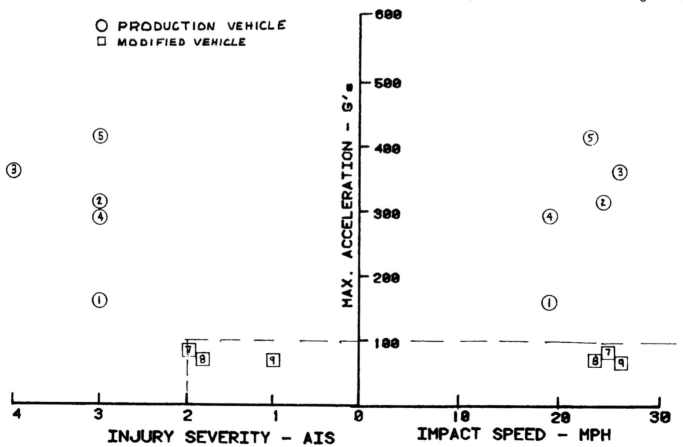

Fig. 1a - Peak knee acceleration versus impact velocity, cadaver data; from Reference 2

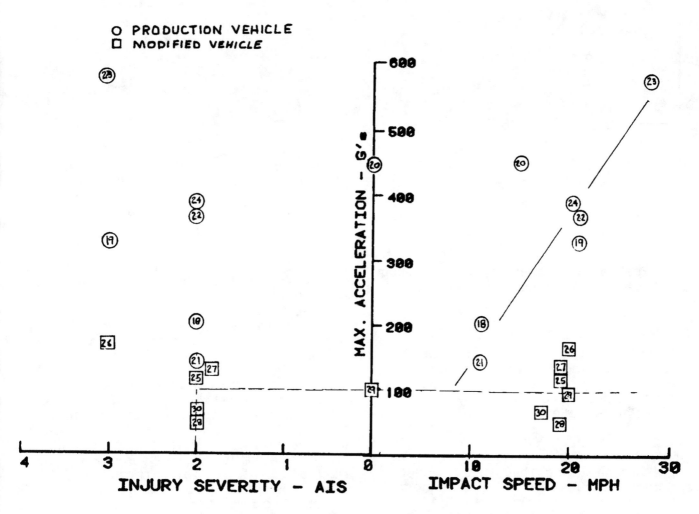

Fig. 1b - Peak knee acceleration versus impact velocity, cadaver data; from Reference 1

pedestrians near the knee region. Injuries include fractures of the knee joint, femur and lower leg long bones and damage to knee ligaments which has potentially more serious consequences than simple fractures to the long bones.

For children, the impacted body regions include the upper legs, pelvis and abdomen. Various child accident victim surrogates have been used in full-scale simulation tests to investigate kinematics and measure impact responses. Battelle Columbus Laboratories (BCL) used a six-year-old child dummy from Humanoid Systems in NHTSA sponsored research tests (2). Child accident simulation tests have been conducted with several baseline models and vehicles modified to reduce pedestrian injury severity. The tests have shown that bumper modifications can substantially reduce the acceleration levels of the child dummy caused by direct impact as shown in Figure 3. There has been some research into injury criteria for children and the correlation between impact response parameters and injury severity (3). Accident data analysis in Reference 3 indicates that children receive AIS 2 leg fractures

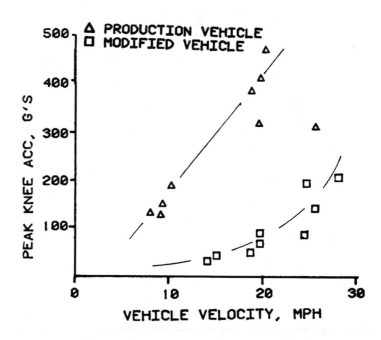

Fig. 2 - Peak knee acceleration versus impact velocity, adult dummy data; from Reference 2

105

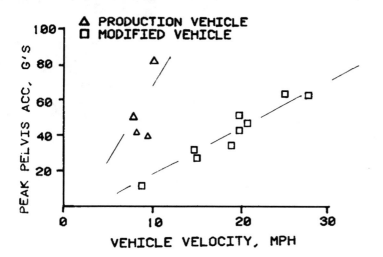

Fig. 3 - Peak pelvis acceleration versus impact velocity 6 year old child dumny data; from Reference 2

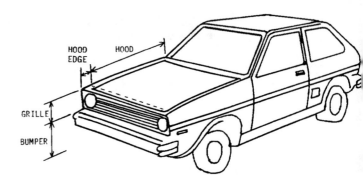

Fig. 4 - Vehicle front surface structure component designation

resulting from bumper contact at a slightly higher speed than adults. This suggests that children would benefit at least as much as adults from bumper modifications designed to limit the leg acceleration response.

PEDESTRIAN - GRILLE/HOOD EDGE INTERACTION - The grille/hood edge area of the exterior surface of passenger cars has large geometric and structural variations from model to model. This area of the vehicle includes the front lights, radiator air inlets, and the leading edge of the hood. For most cars, the height of the grille/hood edge area is between 8 and 15 inches with the lower edge located between 20-23 inches above the ground. The rearward most section of the grille/hood edge is located between 10-18 inches rearward of the vehicle's forwardmost point and is between 28-35 inches above the ground. The design of this area of passenger cars is influenced primarily by styling, aerodynamics, engine cooling requirements and overall vehicle size. Components are generally fabricated from sheet metal, fiber reinforced plastic and polyurethane foam.

On most passenger car designs, structural components are recessed from the exterior surface in this area. The hood edge can roughly be defined as the area in which the vehicle front exterior surface makes the transition from vertical to horizontal. Figure 4 shows the division of the various surface components as they are designated for pedestrian impact research and accident investigation.

The abdomen, upper legs and pelvic regions of adult pedestrian accident victims are likely to contact the grille/hood edge area during a collision with the frontal structure of a passenger car. Consequences of pedestrian impact by this portion of the vehicle frontal structure constitute about 19% of the societal cost of pedestrian accidents.

Adult pedestrian injuries to the upper leg/pelvis body region resulting from direct impact by the grille-hood edge area of passenger cars have severities ranging from AIS 1-4. The injuries are often caused by highly localized forces generated by relatively sharp contours at the vehicle corners and hood edge.

Full-scale accident reconstruction tests have been conducted by researchers to determine the kinematics, acceleration levels and injury characteristics for the body regions struck by this area of passenger cars (1, 2, 8). Cadavers have been used as pedestrian surrogates in these tests as well as anthropomorphic dummies. The pedestrian surrogates used in full-scale accident simulations are generally instrumented with accelerometers mounted on the pelvis. The actual contact area of a standing adult with the grille/hood edge ranges from abdomen to mid-thigh depending on pedestrian height and vehicle geometry with associated variations of the injury type and severity.

In full-scale accident simulation tests the vertical accelerometer position relative to the impact point on the body can vary by 100% or more. An analysis of the BCL (1) impact tests produced a relationship between pedestrian weight, hood edge impact force, and injury severity (4). The analysis indicates that hood edge impact force divided by the weight of the accident victim (Peak F/wt.) has a high correlation with injury severity. Low injury severity levels are likely for low values of Peak F/wt as shown in Figure 5. Figure 6 presents impact response data from full-scale accident simulations using dummy surrogates. This data indicates that some limited reductions in acceleration can be realized. The pelvic impact data from (2) and (8) for dummy and cadaver impact response shows that pelvic impact points on the vehicle and peak pelvic accelerations are similar for the two surrogates.

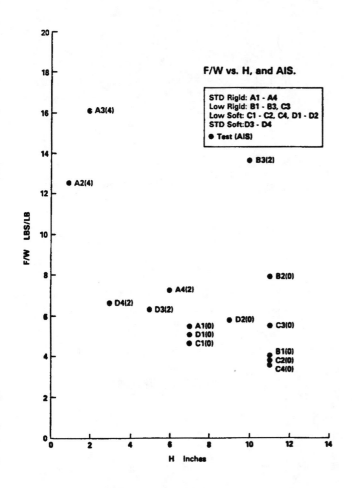

F/W vs. H, and AIS.

STD Rigid: A1 - A4
Low Rigid: B1 - B3, C3
Low Soft: C1 - C2, C4, D1 - D2
STD Soft: D3 - D4
● Test (AIS)

Fig. 5 - Peak pelvis acceleration versus impact velocity - cadaver data; from Reference 2

Tables 2 and 3 indicate that abdominal injuries from grille/hood edge impacts are considerable. Full-scale impact testing with cadavers has not focused on abdominal trauma and data relating impact response parameters to abdominal injury severity is limited.

The child pedestrian may be contacted by grille-hood edge during a collision at all body regions with the exception of the lower leg. A review of the accident statistics and injuries indicates that the head, chest, abdomen, pelvis, and upper legs receive injuries resulting from direct contact with the grille-hood edge area of passenger cars. The high severity injuries resulting from direct contact with this area of the vehicle are primarily to the head, chest and abdomen. They are produced by blunt trauma as well as highly localized forces when sharp contours are involved.

The injury types include skull fractures and concussions, fractured ribs and injuries to the thoracic and abdominal organs. For this reason, design modifications for pedestrian protection in this area must strongly consider the child accident victim. The overall child pedestrian kinematics and impact forces have been investigated through full-scale accident simulations primarily with a 6 year old dummy

surrogate, which is struck at the pelvis and chest by the grille/hood edge. Figures 6 and 7 show the impact response of the body regions contacting the grille/hood edge. The dummy impact response data (Figures 6 and 7) indicates that front structure modifications reduce pelvic impact response significantly, but not the chest response.

PEDESTRIAN - HOOD INTERACTION - The hood of a passenger car can be loosely defined as the entire horizontal frontal exterior surface area located between the hood edge and the windshield, including the horizontal fender surface and the cowl or plenum at the base of the windshield. The hood surface is primarily sheet metal with few, if any, sharp contours. The height at the forwardmost section is between 25 and 35 inches above the ground and 30 to 40 inches above the ground at the rearmost point.

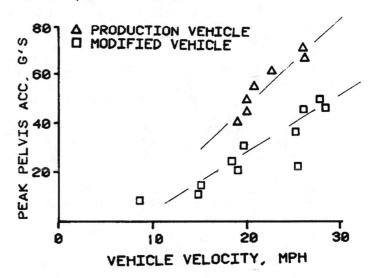

Fig. 6 - Peak pelvis acceleration versus impact velocity - adult dummy data; from Reference 2

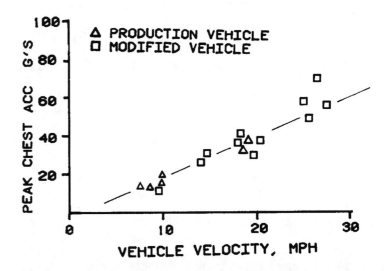

Fig. 7 - Peak chest acceleration versus impact velocity - child dummy data; from Reference 2

Adult pedestrians receive head, chest, upper extremity, pelvis/upper leg and abdomen injuries from direct contact with the hood of a passenger car. The investigation of head impact with the hood represents a major portion of the European pedestrian injury mitigation research although full-scale accident simulations with cadaver surrogates are limited (2, 8). The research indicates that head impact with the cowl area at the base of the windshield and the windshield frame produces high acceleration response and HIC (2000-4000) for impact speeds between 25-30 mph. Also, the head strikes the hood at roughly the initial impact velocity of the vehicle. Tables 2 and 3 indicate that impacts with the hood and fenders produce a high percentage of the pedestrian injuries and these injuries, especially head injuries, contribute substantially to the societal cost of pedestrian accidents.

There is a limited amount of test data available at this time which can be used to develop a relationship between impact response parameters and injury severity for the pedestrian accident victim's head and chest. It can be assumed that the Head Injury Criteria (HIC) and Severity Index (SI), which are used extensively as indicators of vehicle occupant head and chest injury, respectively, can be used to evaluate pedestrian injury. The available cadaver surrogate head impact response data from full-scale impact tests is presented in Figure 8.

Pedestrian head and chest impact response has been investigated through full-scale impact tests using dummy surrogates (2, 6, 8, 9). Some of the available impact test data is presented in Figures 9 and 10. Dummy surrogate impact response is used to determine pedestrian kinematics and to evaluate the relative performance of various points on the vehicle which are contacted in real world accidents. The data indicates that the modification to the hood area mitigates the head and chest impact response.

The head SI for the dummy and cadaver surrogates indicates that the full-scale 50th percentile dummy does not simulate the cadaver head response accurately. Improvements in the upper body for the pedestrian and occupant dummies are currently being considered. Additional work is also needed to determine if more effective modifications are available, and to quantify benefits.

Child accident victims are struck in the head, chest and abdomen regions by the hood of passenger cars. Six year old child dummies, which make head contact with the hood, have been used by researchers to determine kinematics and to evaluate the relative performance of various vehicle designs (2, 3, 6, 14). Results from full-scale impact tests indicate that child acceleration response

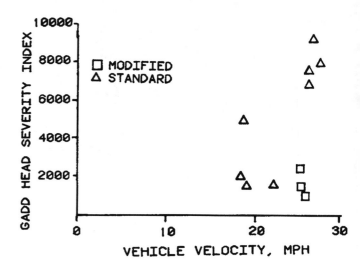

Fig. 8 - Head severity index versus impact velocity - cadaver data; from Reference 2

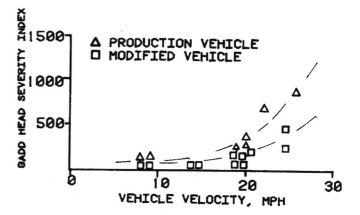

Fig. 9 - Head severity index versus impact velocity - adult dummy data; from Reference 2

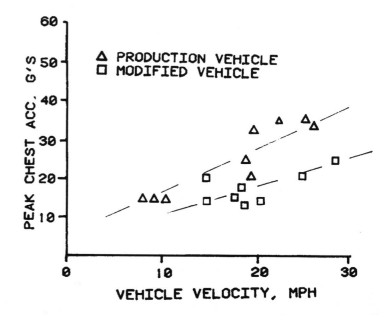

Fig. 10 - Peak chest acceleration versus impact velocity; from Reference 2

levels are higher for children than adults and the child head impact occurs at the forward section of the hood. Biomechanics data relating injury severity to impact response parameters is very limited although the reconstruction of highly documented accident cases has been done and additional work is planned by several researchers, including the NHTSA Safety Research Laboratory.

Currently available information indicates that child head injury tolerance in pedestrian collisions is basically the same as adult tolerance (3). Child dummy surrogate hood impact response data for baseline and modified vehicles is presented in Figure 11. The data indicates that the child dummy head impact response is very similar to the adult response, which is consistent with the accident data findings.

PEDESTRIAN KINEMATICS AND EFFECTS OF VEHICLE FRONT STRUCTURE PROFILE

KINEMATICS - In a typical passenger car collision with an adult pedestrian, the initial contact is between the lower leg and the bumper followed by an upper leg-pelvis or abdominal impact with the grille/hood edge about 20 milliseconds later for a 25 mph collision. The initial impact causes the legs to rotate upward and away from the vehicle while the upper body rotates toward the vehicle until the head and chest strike the hood.

For a 25 mph initial impact speed, the head and chest strike the vehicle about 100 milliseconds after the initial contact with a linear velocity relative to the hood approximately two-thirds of the initial contact speed. The legs continue to rotate upward after the hood and chest have made contact with the hood. If the car is braking by this point, as is usually the case, the car velocity decreases relative to the pedestrian and the pedestrian moves away in front of the vehicle 400-500 milliseconds after the initial contact. The pedestrian continues to rise after breaking

contact with the vehicle and the center of gravity of the pedestrian reaches a height of 20-30 inches above standing height before the pedestrian falls to the roadway 600-700 milliseconds after the initial contact. Schematics of gross pedestrian kinematics are shown in Figure 12.

As shown in Figure 13, the child pedestrian kinematics are slightly different. The initial impact is closer to the child's center of gravity reducing the upward rotation of the legs. The impact with the grille/hood edge is often above the child's center of gravity in the chest or abdomen region causing the victim to be thrown out in front of the vehicle with a very shallow trajectory. Small children (1 to 3 years of age) may be struck high enough by the initial impact to be thrown to the ground rather than into the air. The impact sequence for the child accident victim is shorter than the adult's with the child striking the ground or roadway 300-500

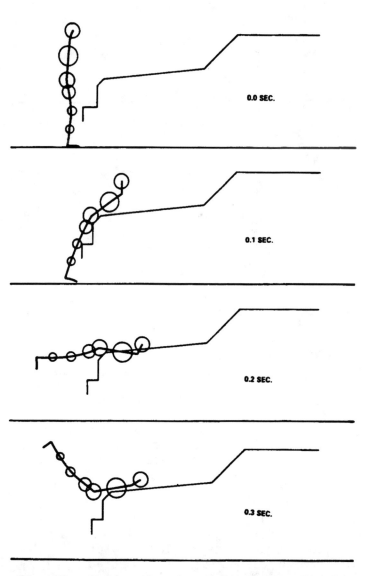

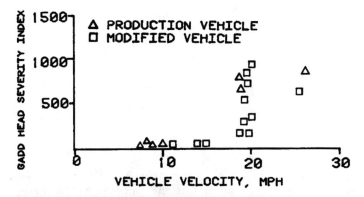

Fig. 11 - Head severity index versus impact velocity - child dummy; from Reference 2

Fig. 12 - Schematic of typical adult pedestrian accident event

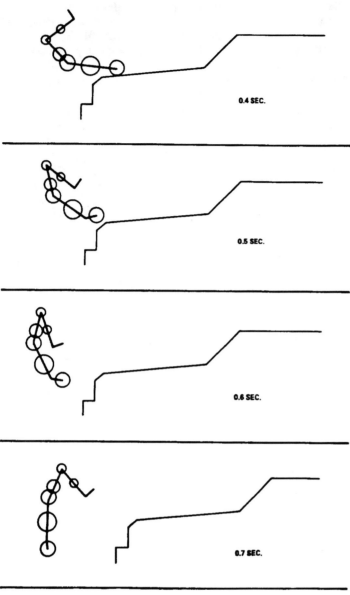

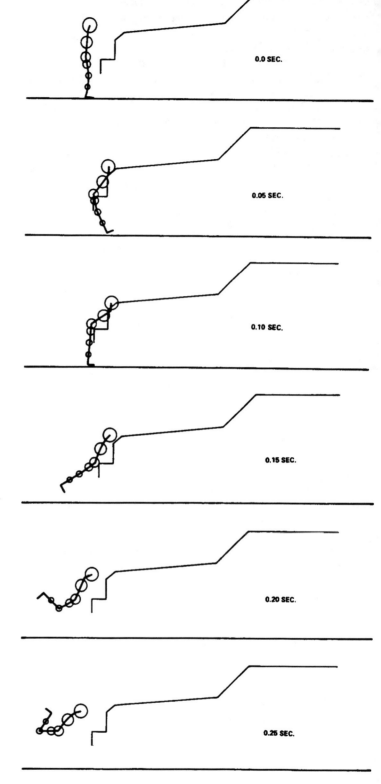

Fig. 12 - Continued

milliseconds after the initial contact for a 25 mph collision.

FRONT STRUCTURE GEOMETRY - The relationship between pedestrian injury severity and the geometry of the front surface structure of the colliding vehicle has been investigated over the past few years by several researchers. (1, 5, 6, 7). Investigation techniques have included accident data analysis, full-scale accident simulations using cadavers and dummies as pedestrian surrogates, and computer based mathematical simulations.

In general, the research has determined that sharp contours at the hood edge and front corners produce higher severity direct contact injuries than large radius contours. The surface profile has an affect on injury patterns and severity with high profile vehicles producing more severe pelvic and

Fig. 13 - Schematic of typical child pedestrian accident event

abdominal injuries and low profile vehicles producing more severe head and chest injuries. The effect of the geometry of a front structure component on injuries to body regions that impact other components and the effect on secondary injuries caused by impact with the roadway must be considered along with the direct contact injury effects.

The effect of vehicle profile geometry on pedestrian injury patterns and injury severity must be evaluated separately with respect to the adult and child accident victim. The contact point of the different body regions on the vehicle is roughly equal to the pedestrian's height, measured from the ground, following the contour of the vehicle's surface. This distance is called the "wrap around" distance. According to accident statistics, head impacts occur at two dominant wrap-around distances, one at slightly under 4 feet and the other at 5 feet, 5 inches, which are the heights of the average sized child and adult accident victim, respectively (9, 16). This indicates that the pedestrian does not slide along the vehicle surface for low and moderate impact speeds. Measuring these distances from the ground up over the frontal structure will define an area across surface structure where head strikes will occur for low and moderate speed collisions (9).

COMPUTER SIMULATION - A matrix of accident simulations was designed to quantify the effects of two vehicle surface profile parameters and bumper stiffness on the adult and child impact response and kinematics. The two geometry parameters investigated are bumper height and hood edge height. Two levels of these parameters were considered representing the extremes for current production passenger car design. Two levels of bumper stiffness were also investigated, one representing current production bumpers and the other designed to reduce lower leg impact response. Thirty-two accident cases were developed and run with a 25 mph impact speed using a computer based math model developed at NHTSA known as MACDAN. The accident simulation matrix is presented in Figure 14 and the math representation of the two-dimensional, six degree of freedom, pedestrian and vehicles used are shown in Figures 15-19.

The results for the 32 accident simulations are presented in Tables 8 and 9. The two levels of the three parameters under investigation are isolated in order to evaluate their effects on pedestrian kinematics and impact response. It can be seen, for example, in Table 8 that the child knee and pelvic accelerations are affected by the variation in bumper stiffness. Comparing Table 8 with Table 9 indicates that the bumper height variation considered has little effect on the responses of the adult pedestrian, but affects the child knee acceleration, pelvis acceleration, and throw height considerably. The hood edge height has opposite affects on the adult and child chest acceleration. The low hood edge produces significantly higher G's for the adult chest than the high hood edge. For children, the effect is reversed.

The data from this matrix of computer accident simulations quantifies the relative impact responses and chest impact velocities for the adult and child pedestrian collisions with vehicles representing the range of production vehicle profile designs. The results agree with the injury patterns observed in accident investigations where attempts were made to correlate injury and vehicle shape (5, 15). That is, high hood vehicles produce higher pelvic and abdominal injury severities than low hood edge vehicles, but lower chest injury severities.

PEDESTRIAN SAFETY ORIENTED VEHICLE MODIFICATIONS

The modification of the front structure of passenger cars to reduce the injury severity of

Bumper Stiffness			Standard				Modified			
Hood Edge Height			High		Low		High		Low	
Hood Length			Long	Short	Long	Short	Long	Short	Long	Short
High Bumper	Child		1	2	3	4	5	6	7	8
	Adult		9	10	11	12	13	14	15	16
Low Bumper	Child		17	18	19	20	21	22	23	24
	Adult		25	26	27	38	29	30	31	32

Fig. 14 - Computer accident simulation case matrix

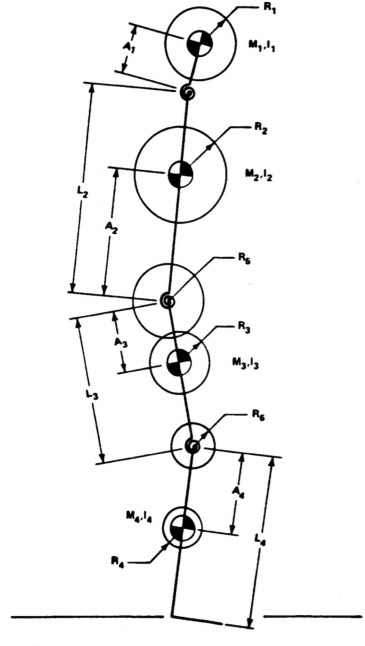

Fig. 15 - Schematic of two-dimensional, 6 DOF pedestrian from Macdan Computer Model

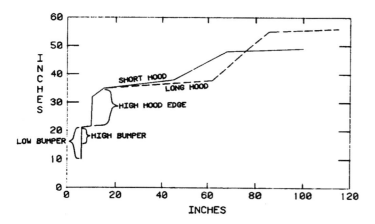

Fig. 16 - High hood edge vehicle used for 32 case accident simulation matrix

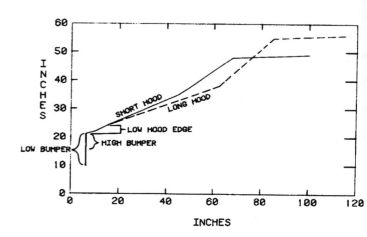

Fig. 17 - Low hood edge vehicle used for 32 case accident simulation matrix

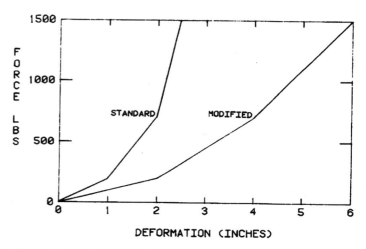

Fig. 18 - Bumper force versus deformation used for 32 case accident simulation matrix

pedestrians struck by them has been the subject of worldwide research for the past five years. Research sponsored by NHTSA has concentrated on modifications to the bumper and grille/hood areas to reduce the severity of direct contact injuries caused by this area of a car (1, 2, 4, 14). Governments and auto manufacturers in Europe have done extensive research into hood and windshield frame modifications as well as bumper and grille/hood edge changes (13, 7, 18). The modifications to production vehicle design involve alterations in the stiffness and geometry of front structure components.

Accident statistics indicate that current front bumper systems produce blunt trauma leg injuries in a high percentage of pedestrian accidents. Bumper stiffness also has an effect on the kinematics and impact severity of other body regions. The accident case matrix highlighted in the previous section indicates that bumper stiffness affects the impact severity of the head, chest and pelvis as well as the direct contact lower leg impact severity for adult pedestrians.

Pedestrian	Vehicle Area		
	Bumper	Hood/Fender Leading Edge	Hood/Fender/Glazing
Adult	Knee/Lower Leg	Pelvis	Head
Child	Pelvis/Upper Leg	Thorax	Head

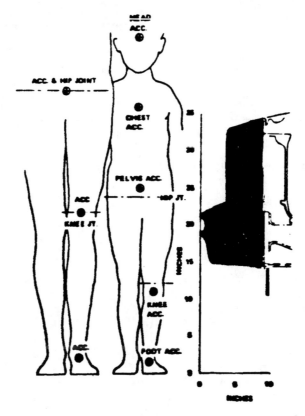

Fig. 19 - Adult and child pedestrians used for 32 case accident simulation matrix

Table 8

Effects of Vehicle Geometry and Bumper Stiffness on Injury Indicators

Child

		Avg. Peak Knee Accel. (G)	Average Peak Pelvis Accel. (G)	Average Chest Impact Velocity Vx Vy Vres (ft/sec)			Avg. Peak Chest Accel (G)	Avg. Peak Head Accel. (G)	Avg. Peak * Pedestrian Throw Ht. (in.)
Bumper Stiffness	Standard	241	142.5	24.6	6.09	26.8	71.1	165.1	26.4
	Modified	138	108.7	26.9	5.60	28.3	64.4	146.6	23.7
Bumper Height	High	157	109.6	25.1	5.10	26.5	69.0	160.5	20.9
	Low	222	141.6	26.3	6.63	28.6	66.5	151.2	29.2
Hood Edge Height	High	179	135.6	35.4	0.64	35.4	77.7	145.9	23.2
	Low	200	115.7	16.1	11.10	19.7	57.9	165.8	26.4

* Indicator of secondary impact severity
 (Roadway)

▣ Significant Effect

113

Table 9

Effects of Vehicle Geometry and Bumper Stiffness on Injury Indicators

Adult

		Avg. Peak Knee Accel. (G)	Average Peak Pelvis Accel. (G)	Average Chest Impact Velocity Vx Vy Vres (ft/sec)			Avg. Peak Chest Accel. (G)	Avg. Peak Head Accel. (G)	Avg. Peak Pedestrian Throw Ht. (in.) *
Bumper Stiffness	Standard	222.6	97.9	11.7	20.5	23.7	47.8	261.9	45.6
	Modified	103.8	82.2	12.7	15.7	20.3	33.6	147.4	39.1
Bumper Height	High	158.8	92.7	10.4	17.2	70.3	37.6	193.5	41.9
	Low	167.5	87.4	14.0	19.0	23.7	43.8	215.8	42.8
Hood Edge Height	High	169.6	114.4	11.7	15.5	19.5	27.8	202.3	39.3
	Low	156.8	65.6	12.7	20.7	24.5	53.6	207.0	45.4

* Indicator of secondary impact severity (Roadway)

▢ Significant Effect

The secondary impact with the roadway may also be affected by bumper stiffness, which influences the throw height of the adult accident victim. With this accident statistics and pedestrian kinematics information, NHTSA sponsored the development of a bumper system designed to limit the impact response of the adult pedestrian lower leg to 100 G's or less for an impact speed of 20 mph or less. The system was installed, along with a modified hood edge, on a 1978 Pontiac Lemans (14, 17).

Another independent bumper, grille/hood edge development program was sponsored by NHTSA in which a Research Safety Vehicle (RSV) built by Calspan Corp. was equipped with a pedestrian protection oriented front structure (18). Both the Lemans and the RSV modifications utilized low density polyurethane foam to fabricate the face bar and high density urethane fascia to form the bumper and hood edge. Dummies were used in full-scale impact tests to evaluate the redesigned Lemans front structure, which showed considerable reductions in impact response for almost all body regions for both the adult and child dummies (14). The results of these tests are shown in Figure 20.

A cost analysis of the Lemans was performed and cost projections for installing the Lemans type front structure modification on several other vehicles were done (20, 21). The cost studies indicated that the Lemans type front end modification could be installed on typical full-size and intermediate cars for approximately the same cost as the production components it would replace. The smaller cars (compact and subcompact models) would incur a cost penalty for installation of the Lemans type front structure because the necessary space was not available. The Lemans type front end has a non-pedestrian benefit which should be considered in a cost/benefit study of the design. The bumper system is substantially less aggressive to other vehicles, especially in low speed side impact collisions (17). The major design features of the Lemans and the Calspan RSV are shown in Figures 21, and 22.

Several European manufacturers have developed Experimental Safety Vehicles (ESV) incorporating pedestrian protection features. A primary consideration was the front structure profile geometry, in which an attempt was made to optimize the trade-off between geometric features that are beneficial to the child and the adult accident victim (7).

The head-hood interaction was also a primary consideration and the cowl, windshield frame and A-pillars were designed and padded to reduce the probability of contact and to mitigate the impact severity (7, 16, 19). The reports on the pedestrian protection oriented design features do not present extensive cost evaluations. Primary cost considerations include the fact that bumpers that meet the requirements of Part 581, The Bumper Standard and provide an improvement in pedestrian impact consequences may be expensive. Also, hood edge/grille design must be consistent with aerodynamic demands so that fuel economy is not sacrificed.

	Vehicle Identification							
	Production		Mod I		Mod II		Mod III	
Bumper foam material	—		16 psi cored		16 psi solid		16 psi solid	
Hood edge foam material	—		16 psi		32 psi		54 psi	
Vehicle velocity, mph	25		25		25		25	
Adult dummy results								
Head—peak acceleration, g's @ ms*	115	109	39	78	49	81	81	109
Head Severity Index	940		385		492		610	
Head Injury Criteria	617		313		385		459	
Chest—peak acceleration, g's @ ms	34	121	43	126	39	122	35	51
Pelvis—peak acceleration, g's @ ms	94	9	78	24	66	21	56	24
Knee—peak acceleration, g's @ ms	350	4	120	24	81	22	60	11
Foot—peak acceleration, g's @ ms	153	8	72	28	106	75	Lost signal	
Six-year-old child dummy results								
Head—peak acceleration, g's @ ms	114	45	131	53	100	55	86	55
Head Severity Index	1090		1200		827		700	
Head Injury Criteria	845		900		579		514	
Chest—peak acceleration, g's @ ms	67	16	96	31	76	26	65	26
Pelvis—peak acceleration, g's @ ms	304	5	147	27	82	24	89	23
Knee—peak acceleration, g's @ ms	367	5	93	28	56	4	54	3
Foot—peak acceleration, g's @ ms	240	11	78	30	78	39	173	39

*MS after initial knee contact.

Fig. 20 - Dummy impact response data; accident simulation tests with modified 1978 Pontiac Lemans (From: Ref. 14)

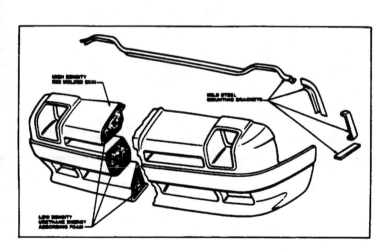

Fig. 21 - Calspan RSV pedestrian impact protection front strucutre (From: Ref. 18)

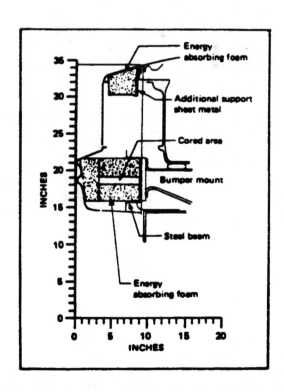

Fig. 22 - BCL modified 1978 Pontiac Lemans; cross section at center (From: Ref. 14)

CONCLUSIONS AND RECOMMENDATIONS

PROBLEM IDENTIFICATION - The PICS file provides a basis for developing statistics on injury frequency per accident and the areas of the vehicle producing the injuries. The PICS file data is not nationally representative of accident frequency by itself but can be combined with other files to produce national estimates.

BIOMECHANICS - The number of full-scale accident simulations using cadavers as pedestrian surrogates is limited, but the data generated to date is useful for establishing the injury severity risk for accidents involving production vehicles. Relationships between injury severity and pedestrian impact response parameters have been examined for most body regions, a major exception being the abdominal region. The 50th percentile adult male dummy needs upper body (head and chest) modifications to more accurately simulate human impact response. Single-degree-of-freedom body segment simulators for the lower leg, pelvis and head have been developed and may be used to evaluate the relative performance of various structures.

Additional full-scale accident tests involving cadavers may be needed to fully assess the injury reduction benefits of front structure modifications. Additional injury tolerance data for children may be necessary to assess potential benefits from vehicle modifications for these accident victims. This could require specially designed accident investigations which provide sufficient data for laboratory reconstruction.

KINEMATICS AND VEHICLE PROFILE GEOMETRY - The relationship between vehicle front structure geometry and pedestrian injury patterns and injury severities has been examined. The pedestrian kinematics (body segment impact velocities and locations) which cause the injury patterns and severities to vary with vehicle front surface geometry have been investigated through full-scale accident simulations utilizing dummies and through analytical methods.

The relationship between vehicle geometry and pedestrian injury patterns is needed for the design of realistic test conditions for body segment simulator use. The front structure geometry may be optimized for a given pedestrian size, but the range of sizes of pedestrian accident victims poses considerable geometric design optimization problems. Front structure geometry must be considered along with modifications in the stiffness of the surface structure material to maximize pedestrian injury reduction benefits.

VEHICLE MODIFICATIONS - The modification of vehicle front structure components for pedestrian injury severity reduction has been investigated by governments and auto manufacturers in the U.S. and Europe. Relative injury reduction benefits for these modified designs have been examined primarily through full-scale impact tests with dummies. The results indicate that significant pedestrian injury severity reductions can be achieved through vehicle front structure modification. Additional accident simulations may be necessary to further assess the beneifts of incorporating these modifications into production vehicle designs.

REFERENCES

1. H. B. Pritz, E. B. Weiss and J. T. Herridge. "Body-Vehicle Interaction: Experimental Study," U.S. DOT Contract No. DOT HS 361-3-745, February 1975.

2. H. B. Pritz, C. R. Hassler, E. B. Weiss. Pedestrian Impact: Baseline and Preliminary Concepts Evaluation," U.S. DOT Contract No. DOT HS 4-00961, May 1978.

3. G. Sturtz. "Biomechanics Data of Children," No. 801313, 24th Stapp Car Crash Conference.

4. R. Eppinger, Y. Kulkarni. "Relationship Between Vehicle Front Structure and Injury to the Adult Pedestrian Hip Area," 8th International Technical Conference on Experimental Safety Vehicles, October 1980.

5. S. J. Ashton, G. M. Mackay. "Car Design for Pedestrian Injury Minimization," 7th International Technical Conference on Experimental Safety Vehicles, June 1979.

6. E. Lucchini and R. Weissner. "Differences Between the Kinematics and Loadings of Impacted Adults and Children; Results from Dummy Tests," 7th International Technical Conference on Experimental Safety Vehicles, June 1979.

7. J. Harris, C. P. Radley. "Safer Cars for Pedestrians," 7th International Technical Conference on Experimental Safety Vehicles, June 1979.

8. A. Heger, H. Appel. "Reconstruction of Pedestrian Accidents with Dummies and Cadavers," 8th International Technical Conference on Experimental Safety Vehicles, October 1980.

9. P. Billault, M. Berthommier. "Pedestrian Safety Improvement Research," 8th International Technical Conference on Experimental Safety Vehicles, October 1980.

10. B. Aldman, et. al. "An Experimental Model System for the Study of Lower Leg and Knee Injuries in Car Pedestrian Accidents," 8th International Technical Conference on Experimental Safety Vehicles," October 1980.

11. "Federal Motor Vehicle Safety Standards: Pedestrian Impact Protection." Federal Register, Vol. 46, No. 14, Janaury 22, 1981.

12. D. Najjar. "Comprehensive Analysis of Motorcycle Accident Data." National Center for Statistics and Analysis, September 1980.

13. M. Kramer. "Improved Pedestrian Protection by Reducing the Severity of Head Impact Onto the Bonnet," 7th International Technical Conference on Experimental Safety Vehicles, June 1979.

14. H. Pritz. "Vehicle Design for Pedestrian Protection," 7th International Technical Conference on Experimental Safety Vehciles, June 1979.

15. K. Langweider, M. Danner, W. Wachter, Th. Hummel. "Patterns of Multi-Traumatisation in Pedestrian Accidents in Relation to Injury Combinations and Car Shape," 8th International Technical Conference on Experimental Safety Vehicles," October 1980.

16. G. Stcherbatcheff. "Pedestrian Protection: Special Features of the Renault E.P.U.R.E." 7th International Technical Confernce on Experimental Safety Vehicles, June 1979.

17. J. T. Herrige, R. D. Vergera. "Initial Evaluation of a Pedestrian-Compatible Bumper System," 8th International Technical Conference on Expermimental Safety Vehicles, October 1980.

18. F. G. Richardson. "Pedestrian Protection nd Damageability and the Calspan Research Safety ehicle," 8th International Technical Conference n Experimental Safety Vehicles, October 1980.

19. J. Z. Delorean Corp. "Cost Evaluation for Four Federal Motor Vehicle Standards-Task VII Cost Review of Pedestrian Safety Modifications," Contract No. DOT HS 7-01767, April 1979.

20. R. McLean, L. Barbarek. "Passenger Car Pedestrian Impact Protection System," Contract No. DOT HS 9-02112, August 1979.

21. A. Malliaris, R. Hitchcock, J. Hedlund. "Setting Priorities in Crash Protection," Presented at the 1982 SAE Congress.

The views expressed in this paper are those of the authors and not necessarily those of the National Highway Traffic Safety Administration.

Benefits from Changes in Vehicle Exterior Design–Field Accident and Experimental Work in Europe

S. J. Ashton and G. M. Mackay
Accident Research Unit
University of Birmingham, England

ABSTRACT

A brief review of pedestrian injury research in Europe is made. The circumstances of pedestrian accidents are described and particular attention is given to the location of the pedestrian's initial contact with the vehicle exterior and to the speed of the vehicle at impact. The dynamics of pedestrian impact are described. Some data on the pattern of pedestrian injury is presented and the general cause of pedestrian injury considered. The influence of bumper height, bonnet height, bumper lead and front structure compliance on pelvic and leg injuries is considered. The influence of vehicle design on vehicle contact head injuries is examined. Consideration is given to the likely benefits to pedestrians that could accrue from changes in vehicle exterior design.

PEDESTRIAN INJURY MITIGATION is a comparitively new concept in crashworthiness. Fifteen years ago little was known about how vehicle design influenced pedestrian injury. Whilst it had been suggested that vehicle design could influence pedestrian injury, and in particular it had been shown that injury potential varied with different vehicle exterior contours, there was and still is confusion as to which shape is preferable. Today there is a considerable body of knowledge about how pedestrians are injured and about how vehicle design influences these injuries. Although there are still areas in which further research is required, sufficient is known to enable vehicles to be designed which are less likely to cause serious injury to pedestrians than vehicles currently on the road. This knowledge has been obtained from studies of real accidents, from the mathematical modelling of the pedestrian impact and from experimental impact tests.

A number of studies, in the mid to late 1960's reported on pedestrian injuries as part of wider studies of accidents (1-4).* However, it was not until the 1970's that pedestrian injury causation was specifically studied. Tharp at the University of Houston reported on 175 accidents involving 190 pedestrians (5). The Transport and Road Research Laboratory in England investigated 149 pedestrian accidents (6) : a particular limitation of this study was a lack of impact speed information.

Major studies of pedestrian accidents were initiated in England and in Germany in the mid-1970's, and in the USA in the late-1970's. The Accident Research Unit at the University of Birmingham began an in-depth study of pedestrain accidents at the end of 1973. By the end of the decade over 700 pedestrian accidents had been investigated, mainly by at-the-scene investigation the research team being alerted by the police or ambulance service within minutes of an accident's occurrence. In Germany in-depth studies of pedestrian accidents have been carried in Hanover since 1973, and more recently in Berlin, under the auspices of the Institute of Automotive Engineering at Berlin Technical University. Although all types of accidents were studied over 200 of these were accidents in which a pedestrian was struck by the front of a car or light van (7,8). In the USA a National Highway Traffic Safety Administration funded project, the PICS study, in which just under 2000 pedestrian accidents were examined has recently been completed (9). Field accident studies have also been conducted in France by both Peugeot-Renault (10) and by ONSER (11).

Complementary to these projects there has more recently been a substantial effort made in Europe to examine in an experimental context the mechanisms of pedestrian injury. This work is reviewed in detail in other papers in the 1983 S.A.E. Symposium, but of particular note are the studies with cadavers by Citroen (12),

---

* Numbers in parentheses designate references at end of paper.

the reproduction of actual pedestrian accidents with both cadavers and dummies under the joint biomechanical research programme, the KOB (13). and the detailed investigation of leg injuries at Chalmers University in Sweden (14, 15).

ACCIDENT CIRCUMSTANCES

LOCATION OF INITIAL CONTACT - The most frequent type of pedestrian accident is that in which a pedestrian is struck by the front of a car or car derivative; approximately 80% of accidents involving cars and between 50 and 60% of all pedestrian accidents being of this type. However, the exact proportion of accidents where the initial contact is to the front of the vehicle varies with age and injury severity. Adults are less likely than children, and elderly adults less likely than other adults, to have a contact with the side of the vehicle. The near-side of the vehicle is more frequently contacted than the offside although again, there are variations with age; adults being less likely than children and elderly adults less likely than other adults to contact the nearside than the offside. For children there are further variations with the age of the child. The younger the child, the more likely is the first contact to be to the nearside of the vehicle (16).

The location of the initial contact influences the severity of the injuries sustained, fatal injuries being more likely to be sustained when the pedestrian is struck by the front of the vehicle than the side. Again age is important - the proportion dying after being struck by the front of a vehicle being greater for elderly adults than for other adults, and greater for adults than for children in general. Very young children are more likely to sustain fatal injuries than older children (16).

The risk of sustaining serious injuries, when struck by the front of a car, varies with the actual location of the contact; initial contact with the outer thirds of the front structure normally results in more serious injuries than contact with the central section. This stems from the fronts of the wings being stiffer than the centre section and from the fact that the subsequent head contact is likely to be on a relatively stiff structure (the 'A' pillar)

ACTION OF PEDESTRIAN - Contact with the nearside of the vehicle occurs almost twice as frequently compared to the offside because a higher proportion of pedestrians are crossing from the nearside when struck. The variations in initial contact location that occur with age may be explained by differences in the way pedestrians of different ages cross the road. Ashton, et al (17), reported that 55 per cent of a sample of 336 pedestrians were crossing from the nearside when struck, compared to only 37 per cent crossing from the offside. The other 8 per cent were either in the road, but not crossing, or on the pavement when struck. It was further noted that children were reported as running when struck in 76 per cent of the child accidents, whilst only 25 per cent of adults and 6 per cent of the elderly adults were reported as running.

IMPACT SPEED - Pedestrian accidents, being essentially urban accidents, tend to occur at comparatively low speeds; approximately 95 per cent of all pedestrian accidents occur at speeds less than 50 kilometres per hour, and about half at speeds less than 20-25 kilometres per hour. A large proportion of these accidents, however, result in only minor injuries. It is generally agreed that, with current car designs, pedestrians struck at impact speeds less than 25 kilometres per hour usually sustain only minor injuries whilst those struck at speeds greater than 30 kilometres per hour usually sustain non-minor injuries. At impact speeds less than 50 kilometres per hour the injuries are likely to be survivable, whilst at speeds greater than 55 kilometres per hour the pedestrians are most likely to be killed. These threshold speeds are for the total population and ignore age effects. The 50 percentile impact speeds for accidents in which pedestrians sustain non-minor injuries is roughly 35 kilometres per hour, and for fatalities alone 50 kilometres per hour.

There are considerable variations in injury severity for a given impact speed, fatalities have been observed at impact speeds less than 20 kilometres per hour and minor injuries at speeds greater than 40 kilometres per hour. These variations in injury severity for a given impact speed indicate that factors other than impact speed are important in determining injury severity.

PEDESTRIAN IMPACT DYNAMICS

For a pedestrian struck by the front of a car or light van, and this is the most common type of pedestrian accident, the first contacts are with the bumper and/or the front edge of the bonnet depending on the shape of the vehicle front structures. The exact location of these contacts on the pedestrian depends on the relative heights of the pedestrian and vehicle front structures. With a young child the bumper will strike the upper leg and the front edge of the bonnet will strike the torso, whereas with an adult the bumper strikes the lower leg and the front edge of the bonnet strikes the upper leg. At low impact speeds (i.e. less than 20 kilometres per hour) these will frequently be the only contacts and there will be little if any damage to the vehicle, the contacts often resulting in only surface cleaning marks to the vehicle.

Above 20 kilometres per hour the pedestrian will angulate and slide over the front edge of the bonnet, the head and upper torso dipping down to strike the vehicle. The exact location of this second contact on the vehicle depends mainly, for current vehicle designs, on the relative height of the pedestrian and the front of the vehicle, the length of the bonnet and the speed of the vehicle at impact. The slip of the pedestrian, relative to the car, increases, for current designs, with increasing impact speed. Children generally strike the top surface of the bonnet with their heads, whereas adult heads strike farther back, often in the windscreen and windscreen frame area. If the impact speed is sufficiently high for there to be a head contact with the bonnet, windscreen or windscreen frame, the impact forces are such that there is normally physical damage to the vehicle.

At very high impact speeds, i.e. above 60 kilometres per hour, an adult pedestrian will generally rotate about this second contact with the vehicle, the body then angulating over the leading edge of the roof, the legs dipping down to strike the top of the roof.

Due to the forces applied to the pedestrian, as a result of the contacts with the vehicle, the pedestrian is accelerated up towards the speed of the vehicle and, should there be braking during the impact (and this is the most common situation) the pedestrian will first attain a common velocity with the vehicle and then, as the vehicle slows at a higher rate than the pedestrian, the pedestrian will travel in advance of the slowing vehicle before coming to rest by sliding and rolling on the ground. If, however, there is no braking during the accident or braking does not occur until a very late stage, the pedestrian may pass over the top, or down the side, of the vehicle and then contact the ground.

Each of these contacts with the vehicle and the contacts with the ground may, and frequently do, cause injury. The initial contact between the front structure and the pedestrian's lower body is commonly referred to as the primary vehicle contact and the head/upper body contact with the vehicle is referred to as the secondary vehicle contact. Should there be contact with the top of the roof by the legs following the head contact then that contact is the tertiary vehicle contact.

PATTERN OF INJURY

Head, leg and arm injuries are the injuries most frequently sustained by pedestrians although the exact frequency, and relative importance, varies with both age and overall injury severity. Multiplicity of injury is the rule in pedestrian accidents. For adults aged 15 - 59 years sustaining only minor injuries the ranking order is legs, head, arms whilst for those with non-minor injuries the ranking order changes to head, legs and arms. The ranking order for survivors with non-minor injuries, counting all injuries, is head, legs, arms pelvis, chest, back, abdomen, neck, whilst for fatalities counting all injuries the order is, head, legs, arms, chest, pelvis, abdomen, back and neck. If only the non-minor injuries are counted the ranking order for fatalities changes to, head, legs, chest, plevis, abdomen, arms, back and neck (Figure 1).

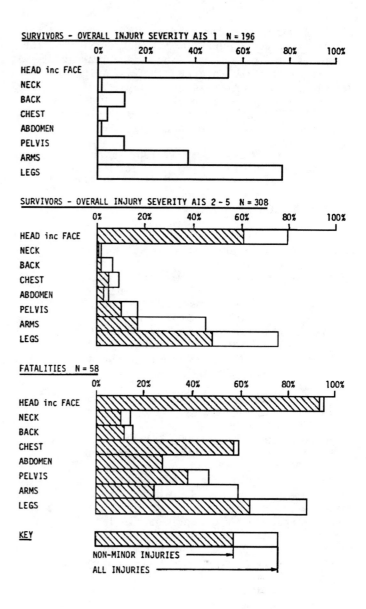

Figure 1 - Pattern of Injury by Overall Injury Severity Counting All Injuries and Counting only Non-minor Injuries for Adults aged 15 - 59 years Struck by the Fronts of Cars or Car Derivatives.

The relative importance of the various body regions is very dependent on the severity of the injuries counted. For example, the ranking order for fatally injured elderly adults if all injuries are counted is head, legs, pelvis, chest, arms, abdomen, beck, neck. If however, only AIS 4-6 injuries are counted the order changes to, head, chest, abdomen, neck, back, legs. (Figure 2).

Ashton et al (18) have reported in detail on the patterns of injury in pedestrian accidents.

GENERAL CAUSE OF INJURY

There is a general concensus from both accident studies and experimental test work that the severity of the injuries resulting from vehicle contact is strongly dependent on the speed of the vehicle at impact but that the severity of the subsequent ground contact injuries is virtually independent of impact speed. At low speeds, i.e. less than 20 km/h, ground contact injuries are frequently the most severe injuries whereas at higher speeds vehicle contact injuries are the most severe injuries. (Figure 3).

PRIMARY VEHICLE CONTACTS

Accident studies (19, 20) have identified the front structures of the car, the bumper and the leading edge of the bonnet, as the main cause of non-minor leg and pelvic injuries sustained by pedestrians. They have also shown that the shape of the vehicle front structure has a significant influence on the location and severity of the injuries sustained. Experimental work (21, 23), has shown that front structure compliance is of equal or greater importance as shape.

Before considering the specific influence that vehicle design can have on leg and pelvic injuries it is instructive to consider the benfits that could accrue from changes in front structure design.

POTENTIAL FOR INJURY REDUCTION - Ashton (24) has considered the effect on the overall injury severity of pedestrians struck by the fronts of cars of designing vehicle front structures such that there are no non-minor leg and pelvic injuries below specified impact speeds. He found that, for a sample of British accidents, the effect of eliminating non-minor vehicle contact leg and pelvic injuries of impact speeds below 40 km/h (24mph) would be a reduction of between 13% and 26% in the number of pedestrians sustaining non-minor injuries.

Interestingly, a similar analysis on a sample of American pedestrian accidents indicated that there would be a reduction of 36% in the number of pedestrians sustaining non-minor injuries if vehicle front structure design was changed such that there were no-minor vehicle contact leg and pelvic injuries at impact speeds

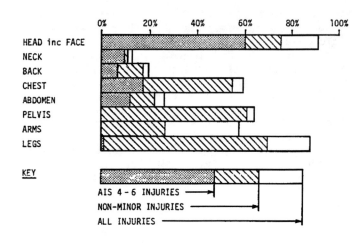

Figure 2 - Pattern of Injury by Severity of Injuries Counted for 127 Fatally Injured Elderly Adults (60+ years) Struck by the Fronts of Cars or Car Derivatives.

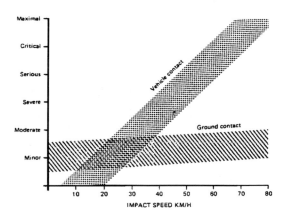

Figure 3 - Predominant Injury Severity by Impact Speed for Vehicle Contact and Ground Contact Injuries Sustained by Pedestrians Struck by the Fronts of Cars or Car Derivatives.

below 25mph. Figure 4 shows the percentage reduction in non-minor casualties resulting from frontal design changes effective up to different impact speeds for the two-environments. These results suggest that the potential for injury reduction by vehicle design is greater for American vehicles than for British vehicles.

These differences may arise from differences in bumper construction between the two sides of the Atlantic. However, other factors, such as exterior vehicle geometry and general structural stiffness may also contribute to the differences in injury potential.

VEHICLE SHAPE - Although the shape of the front structure of a car can be described in a general way, by bumper height, bonnet height and bumper lead, difficulties arise in considering the effects of vehicle shape on pedestrian injuries due to the confounding effects of the pedestrian's height. In order to eliminate the effects of pedestrian height, bumper height and bonnet height can be expressed in the following way:

$$\text{RELATIVE BUMPER HEIGHT} = \frac{\text{ABSOLUTE BUMPER HEIGHT}}{\text{PEDESTRIAN HEIGHT}}$$

$$\text{RELATIVE BONNET HEIGHT} = \frac{\text{ABSOLUTE BONNET HEIGHT}}{\text{PEDESTRIAN HEIGHT}}$$

Bumper lead i.e. the distance by which the bumper is further forward than the bonnet front edge, has an influence on the relative importance of the bumper and front edge of the bonnet as sources of injury. However, this interaction is also influenced by the relative heights of the bumper and bonnet. In order to allow for this effect bumper lead can be expressed in the following way:

$$\text{BUMPER LEAD ANGLE} = \text{TAN}^{-1}\left(\frac{\text{BONNET HEIGHT - BUMPER HEIGHT}}{\text{BUMPER LEAD}}\right)$$

<u>Bumper Height</u> - With current structures, fractures resulting from bumper contact normally occur at the point of contact except when the bumper contact is very low on the leg when, due to the inertia of the leg, the maximum bending moment occurs above the bumper contact. Thus bumper height influences the location of the leg fractures; bumpers located at a relative height less than 0.26 pedestrian height result in lower leg fractures, whilst bumpers located at a relative height greater than 0.31 results in mainly knee and upper leg fractures. Knee injuries are particularly prevalent when the relative bumper height is 0.26 to 0.35 pedestrian height, when a direct contact on the knee is likely to occur. In absolute terms bumpers located at 45 to 54 centimetres above the ground are more likely to result in knee injuries than lower bumpers. Impact forces near the knee joint result in an increased risk of comminuted fractures and damage to the ligaments of the knee joint, and these injuries have serious long-term consequences.

The height of the bumper also has an in-

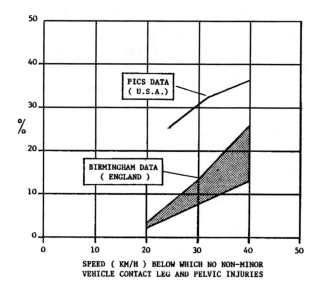

Figure 4 - Percentage Reduction in the Number of Pedestrian Survivors with Non-minor Injuries when Struck by the Front of a Car or Car Derivative if Vehicle Designed such that there were no Non-minor Vehicle Contact Leg and Pelvic Injuries below Specified Impact Speeds.

fluence on the likelihood of a fracture occurring the lower the bumper the less chance there is of the bumper contact causing a fracture. The effective mass of the leg is reduced as the contact height is lowered and the tolerance level may well be different.

Bumpers located at say 35 centimetres above the ground would therefore be preferable to bumpers located at 50 centimetres above the ground, although the optimum height depends on the age, and thus height, distribution of the involved population. This applies only to current relatively stiff narrow bumpers and the introduction of other designs such as full face, compliant front structures may alter this proposal.

<u>Bumper Lead</u> - Bumper lead has a large influence on the relative importance of the bumper and the front edge of the bonnet as sources of injury. As bumper lead increases the importance of the front edge of the bonnet as a source of injury decreases.

Ashton et al (19), reported that when the bumper lead angle is less than 70°, nearly all the leg fractures result from bumper contacts but that as bumper lead angle increases, there is an increase in the number of fractures resulting from contact with the front edge of the bonnet. This is because long bumper leads or more precisely bumper lead angles of less than 70°, result in virtually all the initial contact forces being applied through the bumper.

As the bumper lead angle increases there is early contact with the leading edge of the bonnet as well, the bumper contact forces reducing and the front edge of bonnet contact forces increasing, until a situation is reached where virtually all the contact forces are applied by the front edge of bonnet contact. There is thus an optimum bumper lead angle where the initial contact forces are distributed between the bumper and the front edge of the bonnet, and likelihood of fracture from either is minimised. Vehicles with relatively square fronts appear to be less likely to cause leg fractures than sharp fronted vehicles; accident studies and experimental tests indicate that vehicles with bumper lead angles greater than 80° are less likely to cause leg fractures than vehicles with bumper lead angles less than 70° .

This is because the impact forces are applied at two points along the leg rather than one and this increases the threshold impact speed at which fractures occur. Taking this concept to its conclusion suggests that the vehicle front structure should be smooth, applying a distributed load from mid-calf to pelvis rather than applying two concentrated loads through the bumper and bonnet edge.

Bonnet Height - When there is a contact with the front edge of the bonnet, the height of that edge, as well as the bumper lead angle, influences the pelvic and femoral injuries (20,25). These studies show that pelvic injuries are more likely to occur to pedestrians struck by vehicles with a bonnet height of 75 to 85 centimetres (a relative bonnet height of 0.46 to 0.50 pedestrian height), and that reducing the height of the bonnet lessens the chance of pelvic injury. This is because, as the location of the contact moves away from the pelvis down the leg, the forces on the hip joint reduce.

A recent analysis of pelvic injury in pedestrians suggests that pelvic injuries occur when there is direct loading in the vicinity of the pelvis.(26). It is not the design height of the front of the bonnet that is important but the actual height of the front structure during the impact; frequently the damage to the front structure resulting from pedestrian impact results in an increase in the effective height of the front structure.

VEHICLE COMPLIANCE - Whilst accident studies have shown that, for current designs, the geometric shape of fronts of vehicles has a significant influence on the injuries sustained, experimental testing has indicated that the compliance of the front structures is of greater importance than shape. Reducing the stiffnesses of the bumper and the leading edge of the bonnet has been shown to lower contact forces and thus reduce the likelihood of injury. Concern has been expressed that although increasing the compliance of the front structures will reduce direct contact forces there may be an increase in ground frictional forces due to the compliant front structure holding the foot against the ground for a longer period. To

overcome this Bacon and Wilson (26,27) designed a compliant bumper which lifted the leg off the ground as the bumper deformed.

The main effect of the ground frictional forces is to cause angulation of the lower leg joint this angulation possibly resulting in injury to the knee and ankle joints. Whilst ankle injuries are relatively uncommon in real overcome this Bacon and Wilson (27) designed a compliant bumper which lifted the leg off the accidents, knee injuries are a common type of injury. Ashton (26), reported that knee ligament strain had been found at impact speeds as low as 20 - 35 km/h, but that complete rupture was rare at impact speeds below 40 km/h. Aldman et al (14), however, reported that experimental impact tests on cadaver legs resulted in complete rupture of the knee joint ligaments at speeds as low as 24 km/h. When the leg was struck at the level of the knee joint, Harris (22) suggested that angulation of the knee joint could be controlled by the introduction of a secondary bumper mounted below the main bumper. He recognised that design of the front structure is a compromise and that measures taken to reduce the risk of injury to say the legs of adults may not be the appropriate measures if one is concerned with adult head injuries or injuries to children. He suggested that a good compromise is a vehicle in which the front of the bonnet is located at 700mm above the ground and the bumper is located at 500mm above the ground with a lead of 30mm. He proposed that there should be a secondary bumper located at 300-350mm above the ground positioned below and about 50mm to the rear of the main bumper. This design would limit articulation of the knee to a suitable level, to give a widely distributed loading to the leg and retain some of the advantages that a 500mm high bumper gives to the overall trajectory of the struck pedestrian. The front structure of the vehicle would be designed with consideration of the human tolerance levels and if constructed of a material which deformed under a crash force of 4 KN would require a crush depth of approximately 200m. Such a front end geometry and compliance would diminish greatly the frequency and severity of leg injuries to pedestrians.

SECONDARY VEHICLE CONTACTS

The majority of the injuries resulting from the secondary vehicle contacts are head injuries. Contact with the vehicle has been identified as the main cause of serious head injuries although the exact proportion of serious head injuries resulting from vehicle contact cannot be determined, as in some situations the head injury could have been caused by either the vehicle contact or the subsequent ground contact.

Following the initial contact of the bumper and/or front of the bonnet with the pedestrian's lower body, the pedestrian will either be pushed forward falling to the ground, or bend and slide over the front of the vehicle. The likelihood of a secondary head contact with the vehicle increases with increasing impact speed and decreasing front end height, although above 50 kilometres per hour virtually all pedestrians sustain a head contact with the vehicle.

The location of the head contact on the vehicle is determined by pedestrian height, vehicle front end height and bonnet length, and by the amount of slip of the pedestrian over the front of the bonnet. For current vehicle designs the amount of slip tends to increase with increasing impact speed. The head of an adult struck by an average sized car, dips down to strike the bonnet at moderate impact speeds but at higher speeds the increased slip results in a head contact with the windscreen or windscreen frame. The risk of striking the windscreen or windscreen frame reduces with increasing bonnet length and decreasing pedestrian height.(28).

The severity of the head injuries sustained is influenced by the location of the nature of the head contact on the vehicle and the speed at which the head strikes the vehicle. Accident studies have shown that contact with the windscreen frame is much more likely to result in serious head injuries than contact with the bonnet. Experimental tests have shown that this is because the windscreen frame is considerably stiffer than the bonnet.

The risk of sustaining a non-minor head injury from a vehicle contact can be lessened by either reducing the stiffness of the structure contacted or changing the location of the head contact so that a less stiff area is contacted. The windscreen frame is a difficult structure to modify radically and although it may be possible to change its design so that contacts with it are unlikely to result in non-minor injuries at speeds up to 30 kilometres per hour the benefits that would accrue from this alone would be very small (24). If, however, the location of the head contact could be changed from the windscreen frame to the bonnet greater benefits would be obtainable as it is possible to design the bonnet so that non-minor head injuries are unlikely at speeds below 40 kilometres per hour. It may even be possible to design the bonnet such that head contacts at speeds up to 50 kilometres per hour do not result in non-minor injuries. (29).

The provision of a compliant front structure primarily designed to reduce the likelihood of non-minor leg and pelvic injuries, influences the location of the head contact on the vehicle. The slip of the pedestrian relative to the vehicle becomes virtually zero as a result of the gripping effect of the compliant front structure on the pedestrian's legs. (30,31). This reduction in slip results in the head contact being on the bonnet for all but the smallest cars. The elimination of localised stiff areas on the top surface of the bonnet, such as those that occur at the joins of the bonnet, wings and scuttle, and the provision of suitable under bonnet clearance will minimise the risk of non-minor head injury from

bonnet contact. Thus design of the vehicle exterior can have a significant effect on the severity of the head injuries sustained.

POTENTIAL FOR INJURY REDUCTION - Ashton (24) showed that if vehicles could be designed such that there were no non-minor vehicle contact head injuries at impact speeds below 40 kilometres per hour then there would be a reduction of between 6% and 9% in the number of pedestrians seriously injured. This prediction was based on a sample of accidents involving current designs of European cars.

CONCLUSIONS

The current state of knowledge about pedestrian injury causation is sufficient to show that vehicle design has a major influence on the severity of the injuries sustained by the struck pedestrian.

If vehicles were designed such that there were no non-minor vehicle contact head, pelvic and leg injuries at impact speeds below 40 kilometres per hour (24mph) then there would be a reduction of about one-third in the number of pedestrians seriously injured when struck by the front of a car or car derivative.

Arguably, the benefits from pro-pedestrian car exterior design, for the many countries of the world pedestrian casualties are numerically as important as car occupant casualties, are equal to or greater than the benefits from the provision of passive restraints or the mandated use of active restraints for occupants.

REFERENCES

1. A.J.McLean "The Other Road Users, the Pedestrian, Cyclist and Motorcyclist" Proceedings of the Third Triennial Congress on Medical and Related Aspects of Motor Vehicle Accidents, University of Michigan, Ann Arbor, Michigan. 1969.

2. G.M.Mackay "The Other Road Users" Proceedings of the Thirteenth Conference of the American Association for Automotive Medicine, University of Michigan, Ann Arbor, Michigan, p321, 1971.

3. K.G.Jamieson, A.W.Duggan, J.Tweddell L.I.Pope and V.W.Zvribulis, "Traffic Collisions in Brisbane". Special Report No.2, Australian Road Research Board, Canberra, Australia, 1971.

4. P.M.Culkowski, J.M.Keryeski, R.P. Mason, W.C.Schotz and R.J.Segal, "Research into Impact Protection for Pedestrians and Cyclists". Report No. VJ-2672-V2, Cornell Aernautical Laboratory, Buffalo, New York, 1971.

5. K.J.Tharp "Multidisciplinary Accident Investigation - Pedestrian Involvement" Report No. DOT-HS-801-165. U.S.DOT, Washington, D.C., 1974.

6. T.R.R.L. "Pedestrian Injuries". Leaflet LF317, Transport and Road Research Laboratory, Crowthorne, England, 1974.

7. U.N.Wanderer and H.M.Weber "Field Results of Exact Accident Data Acquisition on Scene". Paper 740568, Society of Automotive Engineers, Warrendale, Pennsylvania, 1974.

8. H.Appel, A.Kuenhel, G.Sturtz and H.Cloeckner "Pedestrian Safety Vehicle, Design Elements, Results of In-Depth Accident Analyses and Simulation". Proceedings of the Twenty-Second Conference of the American Association for Automotive Medicine, Morton Grove, Illinois, Vol II, p132, 1978.

9. J.W.Garrett, A.S.Baum and L.O.Parada "Pedestrian Injury Causation Parameters - Phase II". Report No.DOT-HS-806-148, U.S.DOT, Washington, D.C., 1981.

10. C.Thomas, G.Stcherbatcheff, P.Duclos, C.Tarriere, .J.Y.Foret-Bruno, C.Got and A.Patel "A Synthesis of Data from a Multidisciplinary Survey on Pedestrian Accidents" Proceedings of the Meeting on the Biomechanics of Injury to Pedestrians, Cyclists and Motorcyclists. International Research Committee on Biokinetics of Impacts, Lyon, p351 (A), 1976.

11. M.Ramet and D.Cesari "Bilateral Study-100 Injured Pedestrians -- Connection with the Vehicle". Proceedings of the Conference on the Biomechanics of Injury to Pedestrians Cyclists and Motorcyclists, International Research Committee on Biokinetics of Impacts, Lyon, p102, 1976.

12. P.Billault et al "Pedestrian Safety Improvement Research". Proceedings of the Eighth International Technical Conference on Experimental Safety Vehicles, U.S. DOT, Washington, D.C., p822, 1980.

13. K.H.Lenz. "Joint Biomechanical Research Project-KOB". Unfall-und Sicherheitsforschung Strassenwerkhr, Heft 34, Cologne, 1982.

14. B.Aldman, B.Lundell and L.Thorngren "Physical Simulation of Human Leg-Bumper Impacts" Proceedings of the Fourth International Conference on the Biomechanics of Trauma, International Research Committee on Biomechanics of Impact, Lyon, p232, 1979.

15. B.Aldman, L.Thorngren, O.Bunketorp and B.Romanus "An Experimental Model System for the Study of Lower Leg and Knee Injuries in Car Pedestrian Accidents". Proceedings of the Eighth International Technical Conference on Experimental Safety Vehicles, U.S.DOT, Washington, D.C., p841, 1980.

16. S.J.Ahshton "Some Factors influencing the Injuries Sustained by Child Pedestrians Struck by the Fronts of Cars". Proceedings of the Twenty-Third Stapp Car Crash Conference, Paper 791016, Society of Automotive Engineers, Warrendale, Pennsylvania, p351, 1979.

17. S.J.Ashton, J.B.Pedder and G.M.Mackay "Pedestrian Injuries and The Car Exterior". Paper 770092, SAE Transaction, Society of Automotive Engineers, Warrendale, Pennsylvania.p357 1977a.

18. S.J.Ashton, S.Bimson, and C.Driscoll "Patterns of Injury in Pedestrian Accidents". Proceedings of the Twenty-Third Conference of the American Medical Association. American Association for Automotive Medicine, Morton Grove, Illinois, p185, 1979.

19. S.J.Ashton, J.B.Pedder and G.M.Mackay "Pedestrian Leg Injuries, the Bumper and Other Front Structures". Proceedings of the Third International Conference on Impact Trauma, International Research Committee on Biokinetics of Impacts, Lyon, p33, 1977b

20. S.J.Ashton, J.B.Pedder and G.M.Mackay "Influence of Vehicle Design on Pedestrian Leg Injuries". Proceedings of the Twenty-Second Conference of the American Association for Automotive Medicine, American Association for Automotive Medicine, Morton Grove, Illinois p216, 1978a.

21. H.B.Pritz, C.R.Hassler, J.T.Herridge and E.B.Weis "Experimental Study of Pedestrian Injury Minimization Through Vehicle Design". Proceedings of the Nineteenth Stapp Car Crash Conference, Society of Automotive Engineers, Warrendale, Pennsylvania, p725, 1975.

22. J.Harris "Safer Cars for the Pedestrian". Proceedings of the Conference on Progress Towards Safer Passenger Cars in the United Kingdom, Paper C178/80. Institution of Mechanical Engineers, London, England, 1980.

23. V.J.Jehu, and L.C.Pearson "The Trajectories of Pedestrian Dummies Struck by Cars of Conventional and Modified Frontal Design". Report No.718, Transport and Road Research Laboratory, Crowthorne, England,B9 1976.

24. S.J.Ashton "A Preliminary Assessment of the Potential for Pedestrian Injury Reduction Through Vehicle Design". Proceedings of the Twenty-Fourth Stapp Car Crash Conference, Paper 801315, Society of Automotive Engineers, Warrendale, Pennsylvania, p607, A, 1980.

25. H.Schneider and C.Beier "Experiment and Accident Comparison of Dummy Test Results and Real Pedestrian Accidents". Proceedings of the Eighteenth Stapp Car Crash Conference, Society of Automotive Engineers, Warrendale, Pennsylvania, p29, 1974.

26. S.J.Ashton "Factors Associated with Pelvic and Knee Injuries in Pedestrians Struck by the Fronts of Cars". Proceedings of the Twenty-Fifth Stapp Car Crash Conference, Paper 811026, Society of Automotive Engineers, Warrendale, Pennsylvania, p863, 1981.

27. D.G.C.Bacon and M.R.Wilson "Bumper Characteristics for Improved Pedestrian Safety". Proceedings of the Twentieth Stapp Car Crash Conference, Society of Automotive Engineers, Warrendale, Pennsylvania, p389, 1976.

28. S.J.Ashton, S.Bimson and C.Driscoll "The Influence of Vehicle Size on Pedestrian Head Injuries". Presented at the Fourth International Conference on the Biomechanics of Impact. International Research Committee on Biokinetics of Impacts, Lyon, 1980.

29. M.Kramer "Improved Pedestrian Protection by Reducing the Severity of Head Impact onto the Bonnet". Proceedings of the Seventh International Technical Conference on Experimental Safety Vehicles, U.S. DOT, Washington, D.C., p674, 1979.

30. H.B.Pritz, "Experimental Investigation of Pedestrian Injury Minimization Through Vehicle Design". Paper 770095, Society of Automotive Engineers, Warrendale, Pennsylvania, 1977.

31. J.Harris "Research and Development Towards Protection for Pedestrians Struck by Cars". Proceedings of the Sixth International Technical Conference on Experimental Safety Vehicles, U.S.DOT, Washington, D.C., p724. Also issued as Transport and Road Research Laboratory Report SR238. Transport and Road Research Laboratory, Crowthorne, England. 1976.

Determination of Bumper Styling and Engineering Parameters to Reduce Pedestrian Leg Injuries

Peter J. Schuster (US) and Bradley Staines (UK)
Ford Motor Company

ABSTRACT

The European Commission is proposing legislation aimed at reducing the severity of injuries sustained by pedestrians in the event of an impact with the front-end of a motor vehicle. One aspect of this proposed legislation is reducing the pedestrian's leg injuries due to contact with the bumper and frontal surfaces of a vehicle, assessed using a 'pedestrian leg impact device,' or 'leg-form.'

This proposed legislation presents the challenge of designing a bumper system which achieves the required performance in the leg-form impact—without sacrificing the bumper's primary function of vehicle protection during low-speed impacts. The first step in meeting this challenge is to understand what effects the front-end geometry and stiffness have on the leg-form impact test results. These results will then need to be compared to low-speed impact performance to assess if the two requirements are compatible.

This paper describes an investigation—using concept Finite Element models and a front-end variable geometry vehicle test buck—of the styling and engineering trade-offs for a pedestrian safe bumper system.

INTRODUCTION

Over the past three decades, car manufacturers and legislators have worked diligently to enhance the safety of vehicle occupants. As a direct result of this effort, the number and severity of automotive accidents resulting in injury to the occupants is on the decline.

One area of automotive safety that has received less attention, however, is the protection of pedestrians. While research into pedestrian accidents began in the late 1970's, it was not until recently that considerable effort has been focused on developing a vehicle performance requirement. In 1990, the EC[a] commissioned a group of European automotive safety agencies (TRL[b], INRETS[c], BASt[d], and TNO[e] – the EEVC[f] Working Group 10) to develop a pedestrian impact test procedure that was both repeatable and accurate; replicating a typical pedestrian impact event. The group's original proposals were published in 1991 [1]. These consisted of three sub-system impact test procedures targeted at further reducing the severity of leg, thigh / pelvis, and head injuries (the three most commonly injured areas in a pedestrian impact) at velocities up to 40 km/h (25 mph). The test procedures were proposed as a draft EC Directive in February, 1996 [2]. In addition, these test procedures are being used to evaluate vehicles in the new Euro-NCAP[g] test program sponsored by the U.K. DoT[h], FiA[i], SNRA[j], et al.

The three impact modes presented in the EEVC proposals are (Figure 1):

1. Leg impacts to the vehicle's bumper system and frontal surfaces using a 'free-flight' pedestrian leg impactor (a 'leg-form') [3].

2. Thigh impacts to the vehicle's hood/bonnet leading edge with a guided thigh impact device [4].

---

b. TRL: Transport Research Laboratory (U.K.)
c. INRETS: Institut National de Recherche sur les Transports et Leur Sécurité (National Institute for Transport and Safety Research, France)
d. BASt: Bundesanstalt für Straßenwesen (Federal Highway Research Institute, Germany)
e. TNO: Toegepast Natuurwentenschapppelijk Ondersek (Netherlands Organization for Applied Scientific Research)
f. EEVC: European Experimental Vehicles Committee
g. NCAP: New Car Assessment Program
h. DoT: Department of Transport
i. FiA: Federation Internationale de L'Automobile
j. SNRA: Swedish National Road Administration

---

a. EC: European Commission (provides overall policy direction to each of its 12 member states).

3. Adult and child head impacts to the vehicle's hood-top with two free-flight head impact 'head-forms' [5].

This paper reviews some of the results of an investigation of the styling and engineering implications of the proposed leg impact requirements. In this test procedure, a leg impactor (a detailed discussion of this device is included in the "Leg-Form Impactor" section of this paper) is propelled at a stationary vehicle at a velocity of 40 km/h (approximately 25 mph). The velocity is parallel to the longitudinal axis of the vehicle and can be performed at any point across the front face between the 'vehicle corners.' For this impact event, the proposed performance requirements are:

- Tibia Acceleration (near knee) < 150 g
- Lateral Knee Bend Angle < 15 degrees
- Lateral Knee Shear Deformation < 6 mm

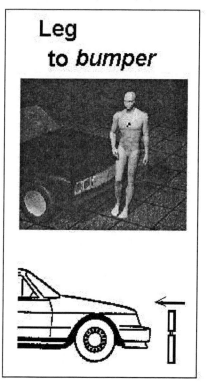
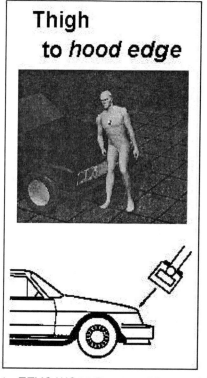
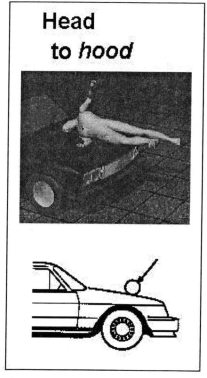

Figure 1. EEVC WG10 Proposed Impact Modes

BACKGROUND

The bumper system has the largest influence on the vehicle's leg impact performance, with the hood leading edge playing a secondary role in limitation of knee bending. Many of the previous papers on this subject are very generic in nature, stating which bumper parameters influence the leg impact performance. In addition, most of the prior work has not used one of the current leg-form impactors.

This earlier work, however, has been essential in the development and implementation of the current test series. In particular, much of the prior work [6-16] has made general recommendations for bumper design which were included in the basic designs tested:

- Lower bumper height-to-ground has been projected to reduce lateral knee bend angle [6,7,8,9,10,11,12], while potentially increasing head impact speed [13].
- A structural lower stiffener [13,14,15] has been proposed as an alternative to a lower bumper height.
- A compliant (soft) bumper system [16] has been used to reduce tibia acceleration, but may reduce vehicle low-speed damage protection.

In order to minimize the influence on (a) the vehicle's styling and (b) the ECE-42 [17] (low-speed damageability) performance, bumper heights should be maintained. Because of this, a structural lower stiffener was added below the existing bumper to reduce lateral knee bend angles. Bumper height-to-ground variation was limited to +/- 25 mm.

Although a very compliant (hollow) bumper system has been shown [16] to perform well in the pedestrian leg impact, this would result in poor performance in the ECE-42 test. Because of this, the following adaptation of a typical bumper system design was chosen as the preferred solution:

- Rigid bumper beam or lower cross-member
- Locally compliant energy-absorbing foam
- Flexible plastic fascia
- Structural lower stiffener

In addition to selection of the bumper system configuration, information on the specific shape and stiffness of these components is also required. The focus of this study was to develop a better understanding of which shape and stiffness characteristics are beneficial to leg-form impact performance.

VARIABLE FRONT-END BUCK

The shape (geometry) and stiffness of a vehicle's front-end are the most significant contributors to pedestrian leg-form impact performance. In order to investigate the specific effects of each characteristic, an adjustable parameterized vehicle front-end design was needed. In particular, the ability to change the bumper (foam) and lower stiffener dimensions, locations, and stiffnesses was required.

To this end, a 'Variable Front-End Buck' which represents the front-end design of a typical European passenger car was developed. It included a bumper, grille, hood/bonnet, and lower stiffener (added below the bumper beam and foam). The buck allowed front-end shape (geometry) and engineering (stiffness) design characteristics to be changed between tests. It represented a 600 mm section across a vehicle front-end (ignoring any curvature). A diagram of the buck (Figure 2) identifies the adjustable geometry and stiffness factors. Table 1 provides a definition of each of the factors.

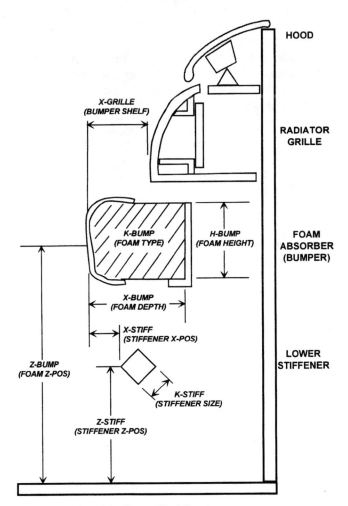

Figure 2. Variable Front-End Buck

The design of the Buck was significantly influenced by the CAE[a] Concept Model development and the results of the CAE DOE[b] (see CAE APPROACH, below). The CAE Concept Model development process identified how and where to attach components. It also indicated that the lower stiffener should have a 'diamond-shaped' cross-section to provide for uniform collapse during the impact. The CAE model also showed that the flexible fascia over the bumper foam influenced the way the foam absorbed energy.

Table 1. Front-End Buck Adjustable Factors

Factor	Description
X-grille	Longitudinal distance from the leading edge of the bumper to the grille.
H-bump	Vertical height of bumper foam.
K-bump	Plateau stress (at 40% deflection) of the PU foam when impacted at 4 km/h.
X-bump	Longitudinal depth of the bumper foam.
Z-bump	Vertical distance from the ground to the center of the bumper foam.
K-stiff	Average load for first 75 mm of stiffener stroke. Related to stiffener size.
X-stiff	Longitudinal distance from the bumper leading edge to the stiffener leading edge.
Z-stiff	Vertical distance from the ground to the center of the stiffener.

The CAE DOE provided an initial indication of which factors were most important to pedestrian leg-form impact performance. These factors were then included in the Buck testing. In addition, the CAE DOE showed that the lower stiffener sizes initially selected (see Table 3) were too far apart (this factor overwhelmed the others in the DOE). Because of this, different sizes were chosen during the Variable Front-End Buck testing (see Table 5).

LEG-FORM IMPACTOR

Pedestrian leg impact performance is assessed through the use of a 'leg-form' impactor. The impactor is constructed from two steel tubular structures (the 'femur' and 'tibia') with prescribed masses, centers of gravity, and moments of inertia. These structures are joined by a knee joint allowing two degrees-of-freedom—'lateral knee bending' and 'lateral knee shear;' hereafter referred to as simply 'bend' and 'shear.' The entire impactor is wrapped in 25 mm of Confor[TM] 'flesh' foam and 6 mm of Neoprene 'skin.'

The characteristics of the knee in shear and bend are specified in terms of quasi-static force-displacement corridors, shown in Figure 3. Note that these tolerance bands are quite wide, especially for quasi-static certifica-

---

a. CAE - Computer-Aided Engineering, including Finite Element Analysis (FEA).
b. DOE - Design of Experiments: A formal process for designing an experiment to get the most information from the least amount of tests. The experimental designs used in this work are more closely associated with Taguchi DOE method than Classical DOE.

tion tests. This is particularly true for bend, where the non-linear relationship between bending load and angle (the metric) exaggerates variability in the measured response. The full certification procedures can be found in the draft regulatory document [2].

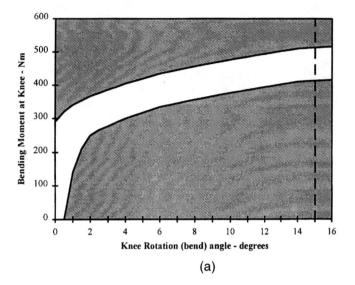

(a)

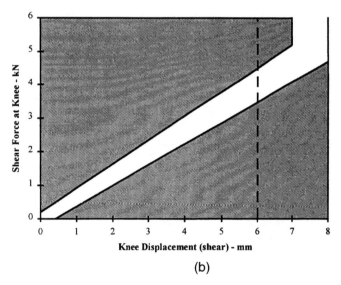

(b)

Figure 3. Knee Certification Corridors

Initial prototype leg-form ligament designs, as proposed by INRETS [3], attempted to satisfy both requirements by using a pair of metal non-linear 'ligaments' able to deform in both bend and shear modes. It was soon noted that this design suffered instability of the ligament when subjected to bend, and metric 'cross-talk'[a] between bend and shear. As a consequence of this, TRL proposed an alternative design in 1995 for the leg-form which separated the bend and shear mechanisms, allowing each to act independently. This solved the instability issues associated with the INRETS design and simultaneously reduced cross-talk [18]. The bend characteristics contin-

ued to be simulated through the use of non-linear ligaments, with the shear compliance achieved through the use of a linear shear spring.

The TRL design has a new concern not seen in the INRETS design—because the shearing displacement is controlled by an elastic spring, the femur and tibia segments can oscillate relative to each other. This 'shear resonance' not only affects the measurement of shear in the knee, but also the acceleration at the top of the tibia segment. TRL is in the process of revising the design to eliminate this concern.

Because of this uncertainty, neither design was used in this investigation. Instead, a MIRA[b]-developed hybrid design, internally known as the 'Simplified Leg-Form,' was used. This has approximately the same mass distribution and bending characteristics as is specified in the EEVC test procedure [1]. However, a shearing mechanism is not included in the design due to the concerns outlined above. It is the opinion of the authors that any system which meets the bend and acceleration requirements would require few changes to also meet the shear requirement.

Comparisons between the mass properties and bending characteristics of the Simplified Leg-Form and the EEVC proposal are shown in Table 2 and Figure 4. While these differences may change the magnitudes of the individual test results, it is the authors' opinion that the trends in the responses will be consistent. Because of this, the bend and acceleration results will only be reported relative to the overall average of the test results.

Table 2.   Leg-Form Mass Properties

	EEVC Proposal		Simplified Leg-form	
	FEMUR	TIBIA	FEMUR	TIBIA
mass (kg)	8.6	4.8	8.2	5.0
I [(1)] (kg-m^2)	0.127	0.120	0.104	0.100
CG [(2)] (mm)	217	233	228	186

1. I: Moment of inertia about the center of gravity
2. CG: Distance from knee center to Center of Gravity

CAE APPROACH

To help shorten product development cycles, a CAE model for the leg impactor and a vehicle modeling methodology have been developed. In addition, these tools were used to determine the initial design for the Variable Front End Buck. All analyses presented in this paper have been performed with RADIOSS[c] version 3.1H or later on a Cray C90.

---

a. Cross-talk: The measurement of one objective datum affects the value obtained for another.

b. MIRA: Motor Industry Research Association (U.K.)

c. RADIOSS: An explicit finite element solver developed by Mecalog (France).

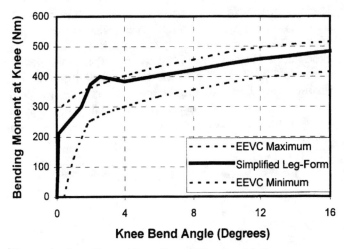

Figure 4. Leg-Form Knee Bend Characteristics

DESCRIPTION OF MODEL – The simplified impactor was modeled with only nine basic parts (Figure 5). They were:

- Femur and Tibia Skins (rubber)
- Femur and Tibia Flesh (foam)
- Femur and Tibia Cores
- Femur and Tibia Rigid Bodies
- Knee Spring

Since this model does not include shear at the knee, a very simple knee model definition was applied. First, the femur and tibia segments were modeled full-length (eliminating the gap between the tibia and femur segments). Knee rotation was then allowed by specifying no interfaces between these two segments in the model. The segments were joined at the center by a zero-length general spring element.

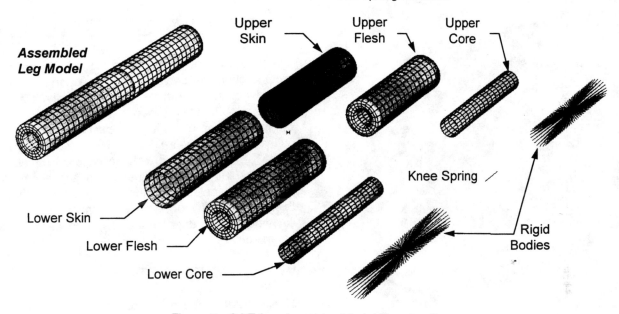

Figure 5. CAE Leg Impactor Model Construction

All degrees-of-freedom for the spring element were constrained with the exception of lateral bending. For this degree-of-freedom, a non-linear function was used to define the bending properties of the knee. Isotropic hardening was used to represent the behavior of the physical knee ligaments, based on the leg impactor static bending certification corridor.

Figure 6 shows the finite element representation of the variable geometry buck. It includes a foam block supported rigidly at its rear face, a bumper fascia to correctly simulate the distribution of force and energy into the foam, and a lower stiffener. In addition, a grill and hood leading edge are included to correctly support the upper portion of the leg during the later stages of the impact.

The Grille, Fascia and Stiffener are all modeled using material Type 2 (elastic-plastic). The foam is modeled using material Type 33 (low density viscoelastic-plastic foam) based on material properties, supplied by Bayer AG, from dynamic crush tests at 4 km/h. The viscous

nature of polyurethane (PU) foams however, means that the properties are often significantly different at higher impact velocities such as the 40 km/hr used in pedestrian leg impact tests. For this reason the supplied data was arbitrary scaled, based on previous high speed PU foam testing experience.

ANALYTICAL PROCEDURE – In order to minimize the number of CAE runs required and to maximize the lessons learned from them, a DOE approach was chosen. Of the eight parameters listed in Table 1, the four deemed to be most significant from previous experience were selected as 'factors' in the DOE. Each of these factors was allowed to take one of three possible values, as shown in Table 3 (Z-stiff was chosen to be dependent on X-stiff in order to maintain a constant approach angle). All other parameters were fixed at levels typically observed on small European cars. For reference, the pedestrian leg-form knee height is defined to be 494 mm from the ground.

The orthogonal array chosen for the DOE was the M27 'probing' matrix. This allows all four of the three-level factors to be used while leaving the main effects and first-order interactions 'clear' (i.e., not confounded with each other).

CAE BUCK RESULTS – A typical sequence of events is illustrated in Figure 7. Maximum tibia acceleration typically occurs between 5 and 10 milliseconds after initial contact with the bumper system. Maximum bending angle typically occurs 10 to 15 milliseconds later.

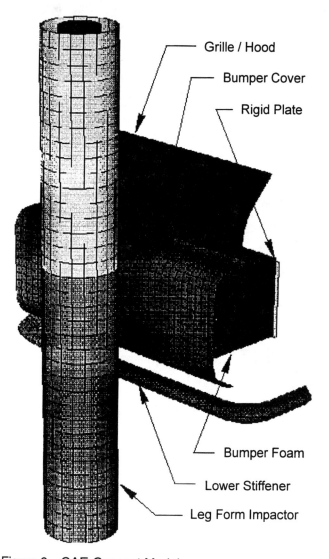

Figure 6. CAE Concept Model

For the DOE analysis to produce valid engineering guidelines, the average of the test results should be near the required target values (from the proposed legislation). The acceleration results of this CAE DOE were well distributed around the 150 G target. However, the knee bend angles were centered around 20 degrees, five degrees higher than the target of 15 degrees. Because of this, the stiffener locations for the subsequent variable buck testing were changed to ensure well-balanced results.

The DOE analysis was performed as an ANOVA (Analysis of Variance) using Minitab. A significance criteria of 90% (1.0 - P > 0.9) was used to evaluate the factors and interactions. This analysis indicated that all four factors were significant relative to the knee bend angle results. However, only two of the factors, K-Bump and K-Stiff, were significant for the acceleration results. In addition, none of the first-order interactions were found to be significant for either of the measured results.

Table 3.   Parameter Levels in CAE DOE

Factor	Levels			Unit
	-1	0	+1	
X-grille	-	65	-	mm
H-bump	-	140	-	mm
K-BUMP	200	250	300	kPa
X-BUMP	70	110	150	mm
Z-bump	-	445	-	mm
K-STIFF	0	1.75	6.25	kN
X-STIFF	-30	-15	0	mm
Z-stiff	265	270	275	mm

The DOE analysis also consisted of viewing main effects plots to check for curvature in the responses and determine whether the ranges selected for the CAE model were appropriate to be used in the physical testing. From the main effects plots, it was observed that the stiffener stiffness (K-Stiff) was linear in both response variables. Also, K-Stiff was found to have opposite effects on the two measured results: Higher spoiler stiffness resulted in lower bend angle, but higher acceleration. Because of this, the K-Stiff factor levels were changed for the physical testing, based on further CAE optimization of this parameter.

VARIABLE BUCK TESTING

The test setup, the experimental design, and the DOE results for the physical test series using the Variable Front-End Buck are presented in this section.

TEST RIG CONFIGURATION – The test setup consisted of the Variable Front-End Buck rigidly mounted to a steel bed-plate placed in front of a Bendix Impactor[a]. There was a carriage attached to the impactor to support the pedestrian leg-form during the initial acceleration of the cylinder. The carriage was stopped after the initial acceleration was complete, allowing the leg to travel the last 0.6 m to the Variable Front-End Buck in free flight at 40 km/h.

---

a. Bendix Impactor - a hydraulic open loop actuator used as a guided mass accelerator to push 9 to 340 kg from 8 to 80 km/h.

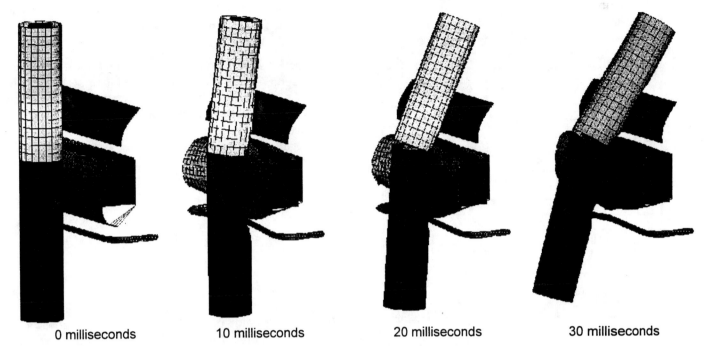

| 0 milliseconds | 10 milliseconds | 20 milliseconds | 30 milliseconds |

Figure 7. CAE Concept Model Animation

A schematic of the Buck was shown in Figure 2. A post-test photograph of the Buck is shown in Figure 8. Sliding attachments and spacer blocks were used for the bumper and stiffener vertical and longitudinal positioning. The plastic grille assembly was attached only at the outboard edges of the Buck, allowing it to bend during the impact. The hood inner panel was attached at the centerline of the buck to simulate a hood latch. The bumper and stiffener components were replaced after each impact. The hood and grille were inspected after each impact and replaced if any structural damage was found.

DESIGN OF EXPERIMENT (DOE) APPROACH – In order to minimize the number of experimental runs, a Screening DOE approach was used. The key questions to be answered by the DOE were:

- Which factors are most critical to the pedestrian leg-form impactor performance?
- Which factors have a non-linear relationship to the responses?
- What are the best settings for the critical factors?

The second question led us to adopt an M18 experimental design, allowing more than two levels for each factor. This design, shown in Table 4, spreads interaction terms across many columns to minimize their influence on a single main effect. Therefore main factor interactions cannot be studied directly with this matrix. This was not a concern for this DOE, since these interactions were found to be weak in magnitude during the CAE Buck analysis.

Figure 8. Variable Front-End Buck

A total of six factors were changed during the testing. This left two columns of the matrix empty to establish the level of noise in the system. Also, one repeat run was performed, to establish the repeatability of the experiment. The six factors and their settings are listed in Table 5. These settings were chosen based on the CAE DOE results. In particular, note that X-Stiff has been extended to move the stiffener in front of the bumper leading edge (+30). Also, K-Stiff was reduced to two levels since its response was found to be linear in the CAE DOE. These two levels were chosen in an attempt to achieve Knee Bend Angle results centered around the target of 15 degrees. While reading this table, recall that the leg-form knee height is 494 mm from ground.

Table 4.   M18 DOE Matrix

RUN	COLUMN							
	1	2	3	4	5	6	7	8
1	-1	-1	-1	-1	-1	-1	-1	-1
2	-1	-1	0	0	0	0	0	0
3	-1	-1	+1	+1	+1	+1	+1	+1
4	-1	0	-1	-1	0	0	+1	+1
5	-1	0	0	0	+1	+1	-1	-1
6	-1	0	+1	+1	-1	-1	0	0
7	-1	+1	-1	0	-1	+1	0	+1
8	-1	+1	0	+1	0	-1	+1	-1
9	-1	+1	+1	-1	+1	0	-1	0
10	+1	-1	-1	+1	+1	0	0	-1
11	+1	-1	0	-1	-1	+1	+1	0
12	+1	-1	+1	0	0	-1	-1	+1
13	+1	0	-1	0	+1	-1	+1	0
14	+1	0	0	+1	-1	0	-1	+1
15	+1	0	+1	-1	0	+1	0	-1
16	+1	+1	-1	+1	0	+1	-1	0
17	+1	+1	0	-1	+1	-1	0	+1
18	+1	+1	+1	0	-1	0	+1	-1

Table 5.   Parameter Levels for Test DOE

Factor	Levels			Unit
	-1	0	+1	
X-grille	-	65	-	Mm
H-bump	-	140	-	mm
K-BUMP	95	125	155	kPa
X-BUMP	70	110	150	mm
Z-BUMP	420	445	470	mm
K-STIFF	1.75	-	4.00	kN
X-STIFF	-30	0	+30	mm
Z-STIFF	240	270	300	mm

ANALYSIS OF RESULTS – The experimental results were analyzed using the Response Surface Model (RSM) method in Minitab version 9.2. Two types of analysis were performed:

- Statistical significance was determined by calculating the coefficient of determination (R^2) for each factor. Significance was defined to be greater than 90%.
- Box-plots[a] were produced to illustrate the effect of each factor on the results.

Table 6.   Significant Factors for Knee Bend Angle

FACTOR	SIGNIFICANCE	LINEAR?
K-Stiff	0.99	YES[1]
X-Stiff	0.99	YES
Z-Bump	0.99	YES
X-Bump	0.94	YES
Z-Stiff	0.93	NO

1. K-Stiff was only tested at two levels. It is assumed to be linear based on the CAE results discussed earlier.

---

a. Box-plot: A plot showing the mean and +/- one standard deviation for each level of a given factor. The mean is shown as a horizontal line and a "box" extends above and below to the standard deviations.

Applicability of the results is limited to the ranges of values which were tested. Some extrapolation is probably acceptable, but caution should be exercised.

Five factors were found to be significant for the maximum Knee Bend Angle. These factors and their statistical significance are listed in Table 6. This table also includes an assessment of whether the response from that factor is essentially linear or non-linear. Figure 9 contains box-plots of the results for each significant factor. To focus on the trends rather than the absolute values, the overall Knee Bend average was subtracted out before plotting.

Table 7.   Significant Factors for Tibia Acceleration

FACTOR	SIGNIFICANCE	LINEAR?
X-Stiff	0.98	YES
X-Bump	0.92	NO

Two factors were found to be significant for the maximum Tibia Acceleration. These factors are listed in Table 7. Figure 10 contains box-plots of the results for these factors. The overall Tibia Acceleration average was subtracted out before plotting.

CAE CORRELATION – A study is currently underway to correlate the CAE Concept Model results to the Variable Front-End Buck test results. Preliminary results from this study have indicated the difference in mass properties (especially tibia C.G.) noted in Table 2 has a significant effect on the knee bend angle results in the CAE Concept Model.

In addition, the correlation study has identified a significant concern with the specification of the knee ligament bending corridor. In an attempt to achieve correlation between the CAE and test results, the knee ligament bending curve used in the CAE model was varied to correspond to (a) the top of the corridor, (b) the bottom of the corridor, and (c) the actual curve generated from the ligaments used in the testing. These changes resulted in knee bend angles which were 10 degrees apart, from 9 to 19 degrees for a single configuration. This variation indicates that the exact bending curve of the knee ligaments used in a test will significantly affect the bend angle results, even if the ligaments fall within the specification.

DISCUSSION

Table 6 and Table 7 indicate that the location and stiffness of the lower stiffener are the most significant of the investigated factors. In addition, the bumper foam height and depth play an important, though lesser, role. These results are in agreement with the recommended bumper designs previously reported [6-16], with the exception of the bumper foam stiffness, which was not found to be significant within the ranges tested in this study. The current work adds quantified results to the previous recommendations—identifying the relative importance of each factor and its effects.

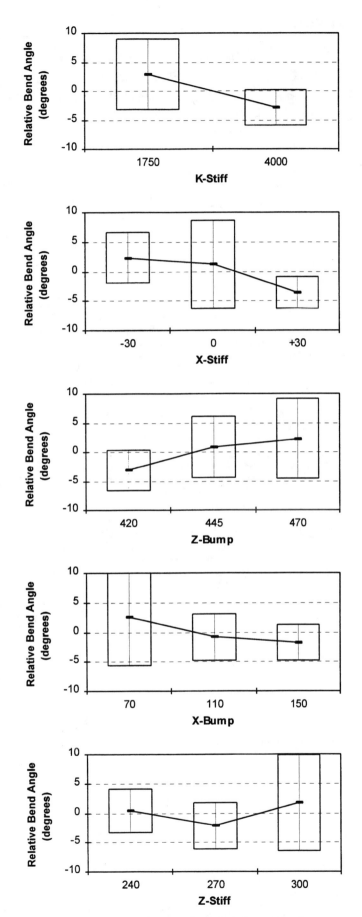

Figure 9. Significant Factors for Knee Bend Angle
(plotted relative to overall test average)

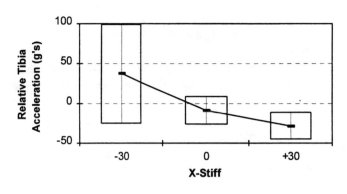

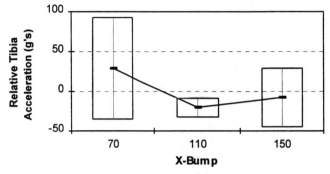

Figure 10. Significant Factors for Tibia Acceleration
(plotted relative to overall test average)

At this point it is important to re-iterate that the results presented are only valid in the ranges tested. In particular, the experimental results suggest that bumper foam stiffness is not a significant parameter in pedestrian leg impact. This observation is only applicable within the stiffness ranges tested (Table 5) – from 0.5 to 1.0 times the stiffness of a typical European bumper energy absorber. The authors believe that the ranges tested for this parameter were too close to identify its critical nature which will likely appear when the stiffness is increased or decreased outside of this range.

As far as the authors are aware, the lower stiffener is a new component not used on current vehicles. Because of this, any issues associated with its addition to the vehicle need to be identified. Several concerns become apparent when considering the addition of this stiff component projecting ahead and below the bumper:

- The stiffener may increase the likelihood of injury to the tibia, fibula, or ankle joint. This possibility has not yet been investigated since the proposed legislation offers no method for measuring ankle or lower tibia injury (the acceleration is measured near the top of the tibia segment on the leg-form).
- The stiffener may increase the velocity of the thigh/ pelvis and head impacts by increasing the speed of the pedestrian's rotation around the vehicle's leading edge.
- High-speed impact performance may be affected, depending on the attachment points and stiffness of the stiffener.

- Low-speed damageability performance will be affected since the stiffener will likely contact some obstacles before the bumper.
- The overall vehicle length will most likely increase, potentially forcing changes to manufacturing plants or shipping operations.
- This study focused on centerline impacts in controlled conditions. Designing a stiffener with the same stiffness characteristics across the entire width of a vehicle remains an open issue.

Three issues associated with the bumper foam depth and position are:

- Deeper bumper foams may affect high-speed impact performance by changing the initial vehicle deceleration seen by the airbag sensor.
- Lower bumper heights may affect ECE-42 performance by moving the bumper system below the specified impact height.
- Deeper bumper foams will result in an increase in the vehicle length, potentially forcing changes to manufacturing plants or shipping operations.

In addition, any of the identified changes to the bumper system will certainly result in increased cost and weight to the vehicle designed to meet the proposed pedestrian leg-form impact requirements.

CONCLUSIONS

There are several styling, packaging, and stiffness factors in the design of a vehicle's front-end which influence pedestrian leg-form impact performance. The focus of this work was to determine which of several selected factors significantly affect the impact test results.

The paper reviewed the development of a standard proposing requirements for pedestrian leg impact. Previously published bumper design recommendations for pedestrian impact were presented, followed by a discussion of issues associated with the two current proposed leg-form impactors.

The methodology utilized a CAE leg impactor model and front-end concept model in addition to a Variable Front-End Buck to investigate the effects of various front-end design parameters on pedestrian leg-form impacts. Six front-end factors were investigated in a DOE using the buck and CAE concept model. The trends identified from the experimental results were found to be consistent with CAE results. In addition, during the CAE correlation, the wide knee ligament certification corridor was found to result in a potentially non-robust measurement of lateral knee bend angle.

The key bumper design factors associated with pedestrian leg-form impact performance were identified. Issues associated with introducing the vehicle front-end design changes suggested by the experimental results were identified for future study.

ACKNOWLEDGEMENTS

The authors gratefully acknowledge the assistance of the following individuals in this work:

- Glenn Klecker, who had the initial idea of a Variable Front End Buck.
- Karin Rothacker, who helped develop the 'Simplified Leg-form' and coordinated the Buck testing.
- Klaus-W. Huland of Bayer AG, who provided the foam samples used in the testing.

REFERENCES

1. Harris, J., Lawrence, G. L., Hardy, B. "Test Methods to Evaluate the Protection Afforded to Pedestrians by Cars." Transport Road Research Laboratories (TRRL) document WP/VS/216, July 1991.
2. "Draft Proposal for a European Parliament and Council Directive Relating to the Protection of Pedestrians and Other Road Users in the Event of a Collision with a Motor Vehicle." European Commission document III/5021/96 EN, February 7, 1996.
3. Cesari, D., Alonzo, F., Matyjewski, M. "Subsystem Test for Pedestrian Lower Leg and Knee Protection." 1991 ESV Conference, Paper # S3-O-08.
4. Lawrence, G.J.L., Hardy, B.J., Harris, J. "Bonnet Leading Edge Sub-System Test for Cars to Assess Protection for Pedestrians." 1991 ESV Conference, Paper # S3-W-22.
5. Glaeser, K.-P. "Development of a Head Impact Test Procedure for Pedestrian Protection." 1991 ESV Conference, Paper # S3-O-07.
6. Cesari, D., Bouquet, R., Caire, Y., Bermond, F. "Protection of Pedestrians Against Leg Injuries." 1994 ESV Conference, Paper # 94-S7-O-02.
7. Sakurai, M., Kobayashi, K., Ono, K., Sasaki, A. "Evaluation Of Pedestrian Protection Test Procedure In Japan-Influence Of Upper Body Mass On Leg Impact Test." 1994 ESV Conference, Paper # 94-S7-O-01.
8. Ishikawa, H., Yamazaki, K., Ono, K., Sasaki, A. "Current Situation of Pedestrian Accidents and Research into Pedestrian Protection in Japan." 1991 ESV Conference, Paper # S3-O-05.
9. Harris, J. "Proposals for Test Methods to Evaluate Pedestrian Protection for Cars." 1991 ESV Conference, Paper # S3-O-06.
10. Aldman, B., Kajzer, J., Anderlind, T., Malmqvist, M., Mellander, H., Turbell, T. "Load Transfer from the Striking Vehicle in Side and Pedestrian Impacts." 1985 ESV Conference.
11. Cesari, D., Cavallero, C., Cassan, F., Moffatt, C. "Interaction Between Human Leg and Car Bumper in Pedestrian Tests." 1988 IRCOBI Conference.
12. Ishikawa, J., Kajzer, J., Ono, K., Sakurai, M. "Simulation of car impact to pedestrian lower extremity: Influence of different car-front shapes and dummy parameters on test results." 1992 IRCOBI Conference.
13. Fowler, J., Harris, J. "Practical Vehicle Design for Pedestrian Protection." 1982 ESV Conference.
14. Dickison, M. "Development Of Passenger Cars To Minimize Pedestrian Injuries." 1996 SAE Conference, Paper # 960098.
15. Kajzer, J., Aldman, B., Mellander, H., Planath, I., Jonasson, K. "Bumper System Evaluation Using an Experimental Pedestrian Dummy." 1989 ESV Con-

ference.

16. Nagatomi, K., Akiyama, A., Kobayashi, T. "Bumper Structure for Pedestrian Protection." 1996 ESV Conference, Paper # xxx.

17. "Regulation # 42: Uniform Provisions Concerning the Approval of Vehicles With Regard to Their Front and Rear Protective Devices (Bumpers, etc.)." United Nations Economic Commission for Europe (ECE) Inland Transport Committee Agreement, March 24, 1980.

18. Faerber, E. "Legform impact tests to evaluate the TRL legform impactor for the EEVC WG10 test procedure for car related pedestrian protection." BASt report, Bergisch Gladbach, December, 1995.

About the Editor

Daniel J. Holt holds a Masters of Science degree in Mechanical Engineering and a Masters of Science degree in Aerospace Engineering. He is currently the Editor-at-Large for SAE's *Automotive Engineering International* magazine. For 18 years Mr. Holt was the Editor-in-Chief of the SAE Magazines Division where he was responsible for the editorial content of *Automotive Engineering International*, *Aerospace Engineering*, *Off-Highway Engineering*, and other SAE magazines.

He has written numerous articles in the area of safety, crash testing, and new vehicle technology.

Prior to joining SAE Mr. Holt was a biomedical engineer working with the Orthopedic Surgery Group at West Virginia University. As a biomedical engineer he was responsible for developing devices to aid orthopedic surgeons and presented a number of papers on crash testing and fracture healing.

Mr. Holt is a member of Sigma Gamma Tau and a charter member of West Virginia University's Academy of Distinguished Alumni in Aerospace Engineering. He is also a member of SAE.

Year 8 Science Workbook

Covers Level Four of *The New Zealand Curriculum*, Science, for Year 8 students.

Raymond Huber

ESA Publications (NZ) Ltd

START RIGHT WORKBOOKS

Year 8 Science Workbook

Raymond Huber

ESA Publications (NZ) Ltd

ISBN 978-1-877530-23-4

(Revised edition 2011)

First published in 2007 by ESA Publications (NZ) Ltd

This edition published in 2011 by ESA Publications (NZ) Ltd

Copyright © Raymond Huber, 2011

Copyright © ESA Publication (NZ) Ltd, 2011

ESA Publications (NZ) Ltd
Box 9453, Newmarket, Auckland 1023, New Zealand

Phone: 09 256 0831
Freephone: 0800 372 266
Fax: 09 256 9412
Email: info@esa.co.nz
Internet: www.esa.co.nz

Editor: Terry Bunn
Compositor: Kalpesh Patel
Cover Design: Jane Meder
Proofreader: Dina Cloete

Printed in India

Contents

Introduction

To the tutor, parents, caregiver or teacher

Year 8 Start Right Science Workbook focuses on *The New Zealand Curriculum,* Science, Level Four. *Year 8 Start Right Science Workbook* is written for Year 8 (11-year-old and 12-year-old) students.

This book aims to stimulate an interest in the world around us and an understanding of the essential concepts of science.

Each Unit introduces and explains a scientific idea. This explanation is followed by experiments, diagrams, questions and puzzles that reinforce the idea. Extensive use is made of New Zealand examples to illustrate concepts. Experiments and tests are designed so that students can do them at home or in school using everyday materials. Where adult supervision is required, this is clearly noted. The Units are written so that most children can work through them independently. (The ability of students varies, so some guidance may be needed.) Class and group activities are also included.

Answers are provided at the back of the book.

Group activities

At the end of each Unit is a group activity which is written for small groups or classes. These activities could easily be adapted for a student to do alone.

© ESA Publications (NZ) Ltd,
Customer freephone: 0800-372 266

The New Zealand Curriculum, Science

The New Zealand Curriculum, Science, is structured around five **strands**.

The Nature of Science strand is the *overarching, unifying strand* and *through it, students learn what science is and how scientists work*. It is made up of the achievement aims Understanding about science, Investigating in science, Communicating in science, and Participating and contributing.

The *contexts in which scientific knowledge has developed and continues to develop* constitute the remaining four Science strands.

- Living World – this strand is *about living things and how they interact with one another and the environment* and is made up of the achievement aims Life processes, Ecology, and Evolution.

- Planet Earth and Beyond – this strand is *about the interconnecting systems and processes of the Earth, the other parts of the solar system, and the Universe beyond* and is made up of the achievement aims Earth systems, Interacting systems, and Astronomical systems.

- Physical World – this strand *provides explanations for a wide range of physical phenomena* and is made up of the achievement aim Physical inquiry and physics concepts.

- Material World – this strand *involves the study of matter and the changes it undergoes* and is made up of the achievement aims Properties and changes of matter, The structure of matter, and Chemistry and society.

Year 8 Start Right Science Workbook is structured around these four Science context strands. The Nature of Science strand is interwoven throughout.

The *New Zealand Curriculum* was implemented in 2010.

Within each of *The New Zealand Curriculum* Science strands there are **Achievement Objectives**.

Nature of Science

Understanding about science

Students will:

- Appreciate that science is a way of explaining the world and that science knowledge changes over time.

- Identify ways in which scientists work together and provide evidence to support their ideas.

Investigating in science

Students will:

- Build on prior experiences, working together to share and examine their own and others' knowledge.

- Ask questions, find evidence, explore simple models, and carry out appropriate investigations to develop simple explanations.

Communicating in science

Students will:

- Begin to use a range of scientific symbols, conventions, and vocabulary.

- Engage with a range of science texts and begin to question the purposes for which these texts are constructed.

Participating and contributing

Students will:

- Use their growing science knowledge when considering issues of concern to them.

- Explore various aspects of an issue and make decisions about possible actions.

Living World

Students will:

Life processes

- Recognise that there are life processes common to all living things and that these occur in different ways.

Ecology

- Explain how living things are suited to their particular habitat and how they respond to environmental changes, both natural and human-induced.

Evolution

- Begin to group plants, animals, and other living things into science-based classifications.

- Explore how the groups of living things we have in the world have changed over long periods of time and appreciate that some living things in New Zealand are quite different from living things in other areas of the world.

Planet Earth and Beyond

Students will:

Earth systems

- Develop an understanding that water, air, rocks and soil, and life forms make up our planet and recognise that these are also Earth's resources.

Interacting systems

- Investigate the water cycle and its effect on climate, landforms, and life.

Astronomical systems

- Investigate the components of the solar system, developing an appreciation of the distances between them.

Physical World

Students will:

Physical inquiry and physics concepts

- Explore, describe, and represent patterns and trends for everyday examples of physical phenomena, such as movement, forces, electricity and magnetism, light, sound, waves, and heat. For example, identify and describe the effect of forces (contact and non-contact) on the motion of objects; identify and describe everyday examples of sources of energy, forms of energy, and energy transformations.

Material World

Students will:

Properties and changes of matter

- Group materials in different ways, based on the observations and measurements of the characteristic chemical and physical properties of a range of different materials.

- Compare chemical and physical changes.

The structure of matter

- Begin to develop an understanding of the particle nature of matter and use this to explain observed changes.

Chemistry and society

- Relate the observed characteristic chemical and physical properties of a range of different materials to technological uses and natural processes.

© Crown copyright

© ESA Publications (NZ) Ltd,
Customer freephone: 0800-372 266

Footers

At the bottom of each page of *Year 8 Start Right Science Workbook* there is a footer which contains the specific **Achievement Objective** (**AO**) or Objectives for the particular Science context strand.

The Nature of Science strand

Space considerations preclude the Nature of Science **Achievement Objectives** appearing as footers on each page of *Year 8 Start Right Science Workbook*. However, the following apply.

AO: 'Appreciate that science is a way of explaining the world and that science knowledge changes over time. Identify ways in which scientists work together and provide evidence to support their ideas.'

Found in:

- Living World Units 2, 3, 5, 7.
- Planet Earth and Beyond Units 1, 5, 6.
- Physical World Units 1, 2, 3, 8.
- Material World Units 1, 2, 8.

AO: 'Build on prior experience, working together to share and examine their own and others' knowledge. Ask questions, find evidence, explore simple models, and carry out appropriate investigations to develop simple explanations.'

Found in:

- Living World Units 1, 2, 5, 6.
- Planet Earth and Beyond Units 1, 2, 4, 7.
- Physical World Units 1, 2, 3, 6.
- Material World Units 1, 2, 5, 6.

AO: 'Begin to use a range of scientific symbols, conventions and vocabulary. Engage with a range of science texts and begin to question the purposes for which these texts are constructed.'

Found in:

- Living World Units 1, 4, 5, 6.
- Planet Earth and Beyond Units 3, 4, 7, 8.
- Physical World Units 2, 3, 4, 7.
- Material World Units 1, 2, 3, 4.

AO: 'Use their growing science knowledge when considering issues of concern to them. Explore various aspects of an issue and make decisions about possible actions.'

Found in:

- Living World Units 2, 3, 5, 7.
- Planet Earth and Beyond Units 4, 6.
- Physical World Units 5, 8.
- Material World Units 6, 7, 8.

The New Zealand Curriculum Key Competencies

The New Zealand Curriculum states that there are five competencies that are the keys to all learning: Thinking; Using language, symbols and text; Managing self; Relating to others; Participating and contributing.

This book encourages the development of these capabilities.

© ESA Publications (NZ) Ltd,
Customer freephone: 0800-372 266

Acknowledgements

All extracts from T*he New Zealand Curriculum* are subject to Crown Copyright and are reproduced in this text by permission of the New Zealand Ministry of Education. Readers are advised to refer to page 18 of *The New Zealand Curriculum*.

My thanks to students at George Street Normal School who tested activities. Also to Terry Bunn for editing and making sure the science was accurate, and to Jane Meder for great illustrations. Thanks to the ESA team for its commitment to quality in educational publishing.

Raymond Huber

Dunedin, July 2011

Image credits

Page 2, Seedlings, Peter Halasz, http://en.wikipedia.org/wiki/File:Monocot_vs_dicot_crop_Pengo.jpg
Page 4, New Zealand sundew plant, Noah Elhardt, http://en.wikipedia.org/wiki/File:Drosera_capensis_bend.JPG
Page 5, New Zealand bush, http://commons.wikimedia.org/wiki/File:Rain_forest_NZ.JP
page 13, *Chlamydomonas*, http://en.wikipedia.org/wiki/File:Chlamydomonas_(10000x).jpg
page 16, *Giardia lamblia*, http://commons.wikimedia.org/wiki/File:Giardia_lamblia.jpg
page 23, Hook grass, *Uncinia*, http://commons.wikimedia.org/wiki/File:Uncinia01c.jpg
page 28, Spiral Galaxy M74, NASA images http://hubblesite.org/gallery/album/pr2007041a/
page 29, Albert Einstein, http://en.wikipedia.org/wiki/File:Albert_Einstein_Head.jpg
Page 31, Cat's Eye nebula, http://www.nasaimages.org
Page 33, The rock planets, http://www.nasaimages.org
Page 34, Surface photograph from the Soviet *Venera 14* spacecraft, NASA images, http://nssdc.gsfc.nasa.gov/photo_gallery/photogallery-venus.html
Page 34, Gas giants, http://en.wikipedia.org/wiki/ Gas_giant
Page 40, San Francisco 1906, http://en.wikipedia.org/wiki/File:San_francisco_1906_earthquake.jpg
Page 41, Allen Carbon
Page 42, Mt Everest, http://en.wikipedia.org/wiki/File:Everestpanoram.jpg
Page 43, Mt Owen, Marlborough, New Zealand, Author photo
Page 48, Hurricane Katrina, http://en.wikipedia.org/wiki/File:Hurricane_Katrina_August_28_2005_NASA.jpg
Page 49, Arun Kulshreshtha, http://en.wikipedia.org/wiki/File:Above_the_Clouds.jpg
Page 55, US Navy freefall parachute jumpers over San Diego, http://commons.wikimedia.org/wiki/File:US_Navy_080722-N-5366K-347_Freefall_parachute_jumpers_leap_from_a_C-130_at_12,500_feet_over_San_Diego_as_part_of_their_regular_training_cycle.jpg
Page 56, Galileo, http://en.wikipedia.org/wiki/File:Justus_Sustermans_-_Portrait_of_Galileo_Galilei,_1636.jpg
Page 56, Leaning Tower of Pisa, Author photo
Page 58, Meteor crater, http://en.wikipedia.org/wiki/File:Meteor.jpg
Page 59, Astronaut Bruce McCandless II, shuttle *Challenger*, NASA images, http://www.nasaimages.org/luna/servlet/detail/NVA2~32~32~65681~128543:Space-Shuttle
Page 61, Supertanker, http://commons.wikimedia.org/wiki/File:Supertanker_AbQaiq.jpg
Page 67, http://en.wikipedia.org/wiki/File:Turbine_aalborg.jpg
Page 68, Solar power system Kennedy Space Center, http://commons.wikimedia.org/wiki/File:Solar_power_system_at_Kennedy_Space_Center.jpg
Page 69, Wairakei Power Station, http://commons.wikimedia.org/wiki/File:Steam_pipelines_towards_Wairakei_geothermal_power_station.jpg
Page 75, US Army mine detection, http://commons.wikimedia.org/wiki/File:Flickr_-_The_U.S._Army_-_mine_detection.jpg
Page 81, Ernest Rutherford, http://commons.wikimedia.org/wiki/File:Ernest_Rutherford_%28Nobel%29.png
Page 104, Salt and pepper, http://commons.wikimedia.org/wiki/File:Spice_4_bg_010104.jpg
Page 105, Hazardous pesticide, http://commons.wikimedia.org/wiki/File:Hazardous-pesticide.jpg

An effort has been made to trace and acknowledge copyright. However, should any omission have occurred, the Publisher apologises and invites copyright holders to contact him.

LIVING WORLD
Evolution

Unit 1   Plant Kingdom

Scientific names

Plants that are similar are grouped together by scientists. Plants are given two names – *genus* and *species*, written in italics or underlined. The genus is the general group and the species is the unique group. For example, the silver beech tree is named *Nothofagus menziesii*. It belongs to a genus of very similar, large trees called *Nothofagus*. Its unique difference is that it has silvery bark when young. The species name is sometimes the name of the person who first identified it. *Nothofagus menziesii* is named after Menzies – a ship's surgeon and plant collector who visited New Zealand in 1791.

The genus name often describes a feature of the plant. For example, Pittosporum, meaning 'pitch seeds', is a group of New Zealand plants that all have sticky seeds.

Activity – Name game

Draw an arrow to match each plant word with the special feature it describes.

Plant word	Feature of plant
trifolium	Deeply serrated leaves
odorata	Very small leaves
serrata	Leaves grouped in threes
microphylla	Plant spreads close to ground
prostrata	Named after Cunningham
cunninghamii	Plant has a smell

Plant classification

Members of the plant kingdom (unlike animals) produce their own food – by photosynthesis. **Plants** can be divided into two groups, according to how they reproduce.

- Some plants reproduce using **spores**. This group includes mosses and ferns and liverworts.

- Some plants reproduce using **seeds**. Seeds mostly come from flowers – 90% of all plants on Earth are **flowering plants**. Seeds can also come from cones – these plants are called **conifers**.

© ESA Publications (NZ) Ltd,
Customer freephone: 0800-372 266

AOs: Begin to group plants, animals, and other living things into science-based classifications. Explain how living things are suited to their particular habitat and how they respond to environmental changes, both natural and human-induced.

Reproducing by seed has made the flowering plants very successful in evolution.

There's such an incredible variety of flowering plants (250 000 species) that accurate classification is important. The two main groups of flowering plant are called monocotyledons (or **monocots**) and dicotyledons (or **dicots**).

Activity – Plant types

Complete the chart:

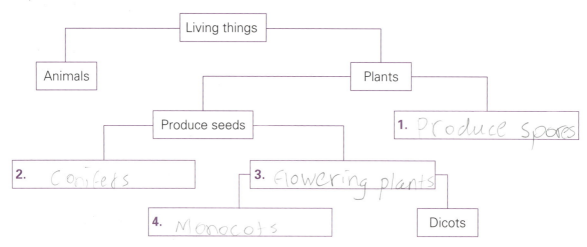

Living things
Animals
Plants
Produce seeds
1. *Produce spores*
2. *Conifers*
3. *Flowering plants*
4. *Monocots*
Dicots

Monocots and dicots

The monocots have only one seed leaf (cotyledon) and leaves with parallel veins. Grasses are monocots – look at the veins running along the length of the leaves. Monocots include palms and orchids.

Dicots have two seed leaves and veins that branch out from a large vein running along the middle of the leaf. Most trees are dicots – look at the network of veins in a pohutukawa leaf. Dicots include roses, beans and cacti.

Activity – Plant classification

Identify each seedling as monocot or dicot:

1. *monocots*

2. *diocots*

1.   2.

AOs: Begin to group plants, animals, and other living things into science-based classifications. Explain how living things are suited to their particular habitat and how they respond to environmental changes, both natural and human-induced.

© ESA Publications (NZ) Ltd,
Customer freephone: 0800-372 266

Flowering plant adaptations

Adaptation means that a plant has evolved a particular size, shape, position, or life cycle to suit the demands of the habitat. Flowering plants have been incredibly successful in adapting to almost every environment on Earth – hot, cold, wet, and dry.

Deserts

Succulents are plants adapted to live in hot, dry places. They can store water in their leaves and stems. Succulents have a waxy coating to prevent evaporation. Cacti are desert plants that have thick, fleshly stems which have ridges so that the stems can rapidly and easily expand when the rains come. They have fast-growing roots which can quickly absorb rainwater (one cactus can absorb 3 000 litres in 10 days). Cacti often have spines instead of leaves, because water is much more easily lost through leaves.

Mountains

Trees in mountain areas have small leaves to reduce heat and water loss (e.g. pine needles).

Alpine plants often have hairs to trap the heat. They grow low on the ground as protection from cold wind. Alpine flowers have bright colours to attract insects – there are fewer insects in cold environments.

Tropics

Rainforest leaves have sharp tips to drain water away. Flowering plants in the tropics often work in partnership with the abundant insects to get pollinated. There are hundreds of different tropical fig trees, each one pollinated by a different kind of wasp. One of the most amazing adaptations is an orchid flower which uses a moth with a 30-cm long tongue to pollinate it.

Group Activity

Investigate how plants are adapted to live in a coastal environment. Take a trip to a beach habitat or estuary and find examples of adaptations. In class, create a chart showing different coastal plant species.

© ESA Publications (NZ) Ltd,
Customer freephone: 0800-372 266

AOs: Begin to group plants, animals, and other living things into science-based classifications. Explain how living things are suited to their particular habitat and how they respond to environmental changes, both natural and human-induced.

Activity – Adapting

Give an example of an adaptation of the following plant parts.

1. Leaves: *Pine needles have thin leaves to reduce water heat loss*
2. Flowers: *Alpine flowers are bright to attract insects*
3. Roots: *cacti have fast growing surface roots*
4. Stem: *Succulents store water in stem*
5. Pollination: *orchids attract moths as polinator*

Shocking science: Meat eaters

Carnivorous plants are adapted to eat animals to get their nutrients (as well as using photosynthesis). They inhabit swampy areas that are low in nutrients such as phosphorus. To compensate for this, they trap insects (and spiders), then digest them to get the nutrients. They use snap traps (Venus flytraps), sticky hairs (sundews), or pitfall traps (pitcher plants). A few trap prey larger than insects.

New Zealand sundew plant

Activity – Research

Research the pitcher plant *Nepenthes rajah*.

How does it trap its food? _____

_____

What does it eat? _____

_____

Unit 2   New Zealand forests

Plant evolution

The first plants on Earth evolved in the sea, then plants gradually colonised the land. The earliest land plants were water-loving plants. Then conifers evolved and used seeds to spread to drier areas. Finally, flowering plants evolved, with their highly-effective seed reproduction.

Colonisation of New Zealand by plants began nearly half a billion years ago.

- Between 443 million years ago (mya) and 417 mya, seaweed-like plants began to live on the land.
- From 354 mya to 290 mya, there were also mosses and ferns appearing and the first conifers grew in forests.
- Between 190 mya and 135 mya, there's evidence of podocarps (e.g. totara) evolving.
- Finally, the flowering plants appeared about 140 mya.

Activity – New Zealand time line

Fill in the missing plants on the time line:

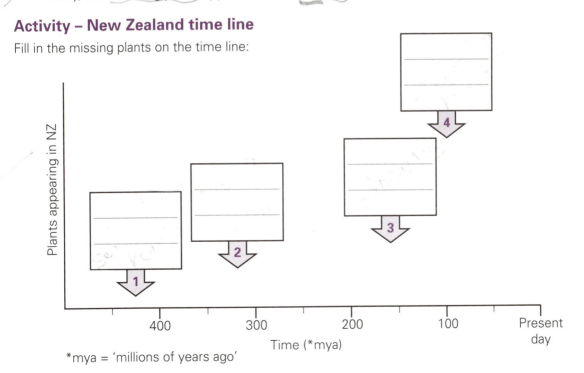

*mya = 'millions of years ago'

© ESA Publications (NZ) Ltd,
Customer freephone: 0800-372 266

AO: Explore how groups of living things in the world have changed over long periods of time and appreciate that some living things in NZ are quite different from those in other areas of the world.

Native forests

The main kinds of forest in New Zealand are **conifer-broadleaf**, and **beech forest**.

Conifer-broadleaf forest

Mostly to be found in the warmer North Island, conifer-broadleaf forests support a wide variety of plants. These forests have a dense canopy – the thick layer of treetops. There are also many climbers and epiphytes (perching plants).

Many of the largest trees are conifers. There are two kinds of conifer – the trees that make cones (e.g. kauri) and the trees that make fruit that encloses the seeds (e.g. totara).

The conifers that produce fruit are called podocarps (meaning 'seed foot') – such as rimu. The seed is enclosed by a fleshy berry like a foot. Podocarps are our tallest and oldest trees. They once covered large areas.

Broadleaf trees such as rata and tawa also grow in these dense forests. They are middle-sized trees, usually smaller than conifers.

Beech forest

Mostly to be found in the cooler South Island, beech forests have a smaller variety of plants. In drier beech forests, it's easy to walk along the forest floor, because there are fewer middle-sized trees. Beech trees can grow at high altitudes, along the Southern Alps. There is now a ban on logging beech forests.

AO: Explore how groups of living things in the world have changed over long periods of time and appreciate that some living things in NZ are quite different from those in other areas of the world.

© ESA Publications (NZ) Ltd,
Customer freephone: 0800-372 266

Activity – Forest species

Name at least two New Zealand examples of each type of forest plant. Write the scientific name (genus and species) and the common name (if it has one).

Conifer
Kauri kaikanaka (mountain cedar

Podocarp
Totara, kahikatea, rimu, matai, miro

Broadleaf
rata, tawa

Beech
silver beech, red beech, mountain beech, black beech!

Climber
supplejack, clematis, jasmine

© ESA Publications (NZ) Ltd,
Customer freephone: 0800-372 266

AO: Explore how groups of living things in the world have changed over long periods of time and appreciate that some living things in NZ are quite different from those in other areas of the world.

Save our forests

Maori and European settlers have cut down and burned 90% of New Zealand's native forests. European settlers, especially, also introduced animal pests and weeds which can seriously affect or sometimes even destroy forests. Native forests are absolutely essential for New Zealand's future, providing:

- soil protection – trees protect the soil from heavy rain, erosion, and flooding
- gas exchange – trees produce oxygen, and absorb CO_2, so they reduce global warming
- habitats – forests are the habitat for many native birds and insects
- water cycling – forests cycle water between the soil, plants and the atmosphere
- recreation – forests are a major tourist focus.

Activity – Natural processes

1. Explain how forests reduce soil erosion.

_____

_____

2. Explain how trees absorb CO_2.

_____

_____

3. Explain why birds need forests.

_____

_____

How can you support native forests?

- Join a local conservation group, such as Forest and Bird.
- Plant native trees.
- Stay on tracks when bush walking.
- Clean up weeds.

Group Activity

The group plants a small area with native plants or trees, choosing species that suit the area, or 'adopts' an area of existing bush to care for.

AO: Explore how groups of living things in the world have changed over long periods of time and appreciate that some living things in NZ are quite different from those in other areas of the world.

Unit 3   Bird evolution

Archaeopteryx – *the early bird*

Birds are classified as vertebrates – animals with an internal skeleton. Birds evolved from dinosaurs about 150 million years ago. There were gliding reptiles called pterosaurs, but they had membranes instead of feathers. The 'missing link' between reptiles and birds was a small fossilised animal found in 1861. The fossil skeletons of *Archaeopteryx* have a reptile's teeth and tail, but also bird-like wings and legs. After the dinosaurs suddenly disappeared, there was a huge increase in the number of bird species. Today there are about 10 000 different species of bird.

Birds have evolved body features that have adapted them for powered flight.

Archaeopteryx fossil

Flight adaptations

Bones

Birds have strong, but light, skeletons – many of the bones are hollow to reduce weight. (Snap a dry chicken bone to check this.) The skeleton of a seagull is about 5% of its total body weight – compare this with humans' 12–15%.

Birds also have large breastbones (the keel) to anchor the wing muscles. Birds have strong, lightweight beaks, instead of heavy jaws with teeth.

Wings

Wings must be strong, flexible and able to trap air. Wing bones are typical of vertebrate forelimbs, and can bend like levers. Wings are curved on top to create lift. The evolution of feathers made the bird unique among all other animals. Feathers are used to steer and to lift, and create a solid surface to push against air.

Body

Birds have evolved a streamlined shape to reduce air resistance. They have a high metabolic rate – meaning they can produce energy quickly enough for the high energy demands of flight.

The tail assists in changing direction in flight.

AO: Explore how groups of living things in the world have changed over long periods of time and appreciate that some living things in NZ are quite different from those in other areas of the world.

Activity – Seagull

Label the picture to show six ways in which a bird is adapted for flight:

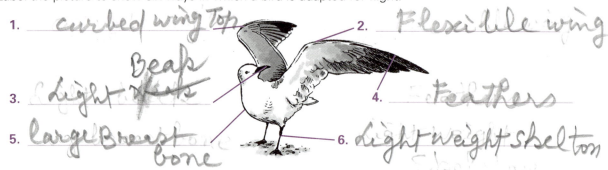

1. *curved wing Top*
2. *Flexible wing*
3. *Light* ~~Hats~~ *Beak*
4. *feathers*
5. *large Breast bone*
6. *Light weight skelton*

Activity – Wings and arms

Human forelimbs are called arms. Wings are a bird's forelimbs. Birds' wings and human arms have similar evolutionary origins. The wing has three digits, a human arm has five. Label the bones of the human arm using the same labels as the bird's wing.

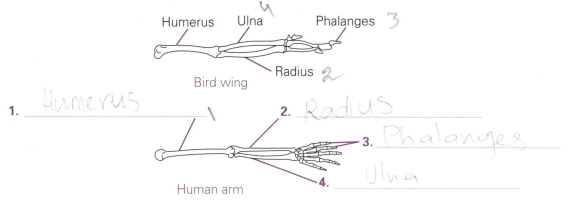

Humerus   Ulna *4*   Phalanges *3*

Radius *2*

Bird wing

1. *Humerus*
2. *Radius*
3. *Phalanges*
4. *Ulna*

Human arm

Beak adaptations

Birds use their forelimbs for flight, so their beaks have adapted to hold food. (There are exceptions such as birds of prey, which also use their feet to seize food.) Birds' beaks have evolved into an amazing variety of different shapes, mainly because of the huge variety of foods they eat. Bird diets include meat, fish, plants, fruit, seeds and nectar.

- Seed eaters (such as sparrows) have evolved short, thick beaks for cracking seeds.
- Meat eaters (such as hawks) have sharp, curved beaks for grabbing and cutting flesh.
- Nectar eaters (such as hummingbirds) have long narrow beaks for probing deep into flowers.
- Woodpeckers have sharp, chisel-like beaks for chipping holes in wood to eat insects.
- Ducks have broad, flat beaks for straining small plants and animals from water.
- Parrots eat very hard nuts and seeds, so they have thick, powerful beaks.
- Fishing birds (such as herons) have long, spear-like beaks for catching prey.
- Pelicans have huge beaks that scoop up large volumes of water containing fish prey.
- Crows have an all-purpose beak, because they eat a varied diet. Blackbirds are versatile too – they have a sharp beak for picking up seeds, but it's also long enough to grab worms.

© ESA Publications (NZ) Ltd, Customer freephone: 0800-372 266

Activity – Beak shapes

Find an example of each beak and draw the beak in the *Shape* column:

Beak type	Adaptation	Shape
Cracker	Cracking seeds	
Cutter	Tearing meat	
Strainer	Straining water	
Nutcracker	Crack hard nuts	
Spear-like	Catch fish	
All-purpose	Varied diet	

© ESA Publications (NZ) Ltd,
Customer freephone: 0800-372 266

AO: Explore how groups of living things in the world have changed over long periods of time and appreciate that some living things in NZ are quite different from those in other areas of the world.

Extreme science: Huia beaks

Both Maori and Europeans used huia feathers for decoration, and the huia was unfortunately wiped out by 1907. The huia had extreme beak specialisation – the male and female having completely different-shaped beaks.

- The male's short strong beak was used to poke holes in trees to find insects.
- The female's beak was long and curved for picking out insects from the holes, and for feeding them to the male.

Activity – Draw a beak

Find a picture of huia and draw the beaks here:

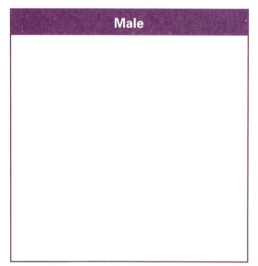

Male

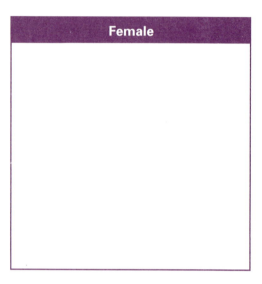

Female

Science history: Darwin's theory

Charles Darwin's ground-breaking theory (1859) described how plants and animals evolved over a long time. His idea is sometimes called 'the survival of the fittest'. The 'fittest' are animals that have a genetic advantage which makes them better able to get resources, avoid predators or find mates. They are likely to have more offspring which survive, and inherit the advantages.

Why has Darwin's theory been controversial? _____

_____

_____

Group Activity

Science knowledge changes over time. Find out how ideas about evolution have developed over the last few hundred years. Discuss how these changes have influenced people's beliefs and culture.

AO: Explore how groups of living things in the world have changed over long periods of time and appreciate that some living things in NZ are quite different from those in other areas of the world.

© ESA Publications (NZ) Ltd,
Customer freephone: 0800-372 266

Ecology

Unit 4   Protists

Protist Kingdom

The Protist Kingdom is *mostly* single-celled organisms – algae and protozoa. Protists are essential to life on Earth, but some can cause serious human diseases.

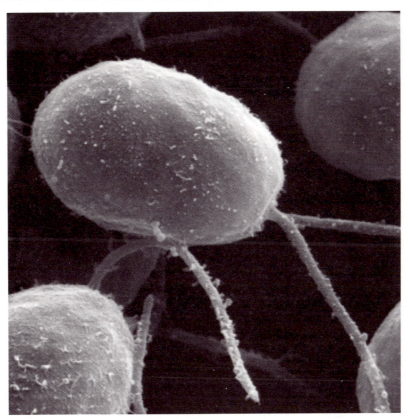

Chlamydomonas

Chlamydomonas is a single-celled green alga that has a 'tail' that can be whipped around to enable movement.

Algae

Algae are plant-like protists which use photosynthesis (sunlight) to get their energy. They live mainly in water. Algae can be green, brown or red. Most of the plants living in the sea are kinds of algae. They can grow as deep down as light can reach.

The basis of almost all ocean food chains are the algae called phytoplankton. Phytoplankton can float in dense masses so large they can be seen from space. They produce 50% of the Earth's oxygen!

AO: Explain how living things are suited to their particular habitat and how they respond to environmental changes, both natural and human-induced.

- **Green algae** – freshwater ponds can have large amounts of green algae floating like a blanket of slime. Green algae are an important food source for pond animals (e.g. tadpoles and snails).
- **Diatoms** – microscopic, single-celled algae. The cell is protected inside a beautiful case made of silica (a mineral used to make glass).
- **Seaweed** – multi-celled algae found near the coast. Seaweeds have no true roots and no tissues for conducting water. Seaweeds have adaptations for living in the rough ocean, including air bladders or spongy cells for flotation, flexible stems, mucus-covered leaves to prevent drying out in the air, and a holdfast (foot) to anchor onto rocks.

Seaweeds provide food for fish, shellfish, and kina. Humans eat seaweeds too (there's only *one* poisonous variety in New Zealand) – they are a good source of vitamins. New Zealand has 850 native seaweeds, including sea lettuce, Neptune's necklace, and karengo (related to Japanese nori – used in sushi). The largest are the kelps, which grow up to 20 m long. Some seaweeds contain jelly-forming chemicals which are extracted and used in yoghurts, shampoos, toothpaste and ice cream.

Activity – True or False?

Circle the correct answer.

1. Algae are plants because they photosynthesise.   True / **False**
2. Algae need light.   **True** / False
3. Algae are all single-celled.   True / **False**
4. Diatom cells are made of silica.   **True** / False
5. Mucus stops seaweed drying out at low tide.   **True** / False
6. Seaweed stems are rigid to protect them from waves.   True / **False**

Protozoa

Protozoa are animal-like protists which get their energy from other organisms. Protozoa are mostly predators that eat algae and bacteria, but some are parasites. Although they can move, they are *not* classified as animals.

Movement

Protozoa are grouped according to how they move.

- Ciliates are protozoa that move by beating cilia – tiny, hair-like projections.
- Flagellates have long, whip-like tails called flagella (singular – flagellum), which they use to propel themselves.
- Amoebae are 'blobs' that can change their shape and stretch themselves out to move.

Diseases

Some protozoa are parasites that absorb food from a living host. They are spread by insects and cause diseases such as malaria and sleeping sickness. Malaria kills up to 2 million people a year in the tropics.

Activity – The malaria cycle

Malaria parasites are protozoa called *Plasmodium*. *Plasmodium* infects humans through mosquitoes. A mosquito infected with *Plasmodium* injects it into the bloodstream of a person; *Plasmodium*

© ESA Publications (NZ) Ltd,
Customer freephone: 0800-372 266

invades the red blood cells and multiplies; the cells burst and release more parasites into the person's blood; a *Plasmodium* is sucked up by another mosquito; it multiplies inside the mosquito; the cycle continues when the mosquito bites another person.

Label the diagram using the following letters:

A   *Plasmodium* multiplies in mosquito.

B   *Plasmodium* released into human's bloodstream.

C   *Plasmodium* injected into human's bloodstream.

D   *Plasmodium* in human blood sucked up by another mosquito.

E   *Plasmodium* multiplies in person's red blood cells.

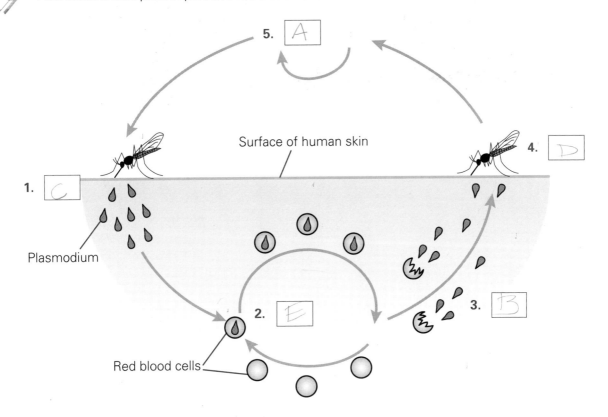

Group Activity

Interpreting diagrams

Look at the diagram of the malaria cycle (above) and discuss the following questions with the group.

* What do the arrows represent?

* What exactly does this diagram tell us? Make a list of facts.

* What does this diagram not tell us? For example, the diagram does not tell us the sex of the mosquitoes shown.

* How could additional information be shown on the diagram?

* Would photos be more effective?

Shocking science: Extreme tummy bug

Not all parasitic protozoa are spread by insects. *Giardia* gets into some New Zealand waterways from animal faeces. People drink the water and the *Giardia* parasite attaches to their intestine wall, causing excessive diarrhoea and farting. Fortunately, the parasite can be removed with medicine.

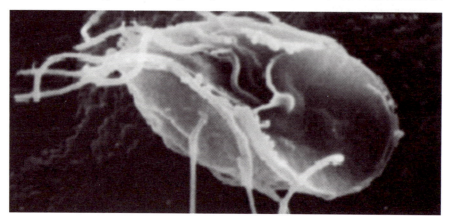

Giardia lamblia

AO: Explain how living things are suited to their particular habitat and how they respond to environmental changes, both natural and human-induced.

*© ESA Publications (NZ) Ltd,
Customer freephone: 0800-372 266*

Unit 5   Viruses

Alien invaders

Viruses may be the strangest microbes on Earth; they are certainly the most deadly. 'Virus' is a Latin word meaning *poison*. The smallpox virus alone killed and disabled hundreds of millions of people throughout history – until it was finally wiped out in the 1970s.

There's always been debate about whether viruses can be classified as living things. Viruses are small chemical 'packages' which cannot feed, grow, or reproduce independently. Sounds harmless? When a virus invades a living cell (of a **host**), it suddenly becomes very active. It takes control of the host cell and makes copies of itself – copies that quickly become an army.

Viruses have two main parts – a core of genetic instructions (viral **genes**), surrounded by a protective coat called a **capsid**. The capsid is often covered in spikes that help the virus attach to a host cell. Viruses are extremely small – the virus that causes the common cold could fit half a billion copies on the head of a pin! They range from 40 to 400 nanometres in diameter.

Viruses can invade the cells of animals, plants, and micro-organisms. They need to be transported to hosts in water droplets, air, food, or blood.

Humans have become more vulnerable to viral attack due to urbanisation and travel. Viruses cause diseases such as influenza (flu), hepatitis, AIDS, herpes, and Ebola.

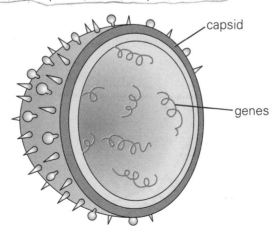

Flu virus

AO: Explain how living things are suited to their particular habitat and how they respond to environmental changes, both natural and human-induced.

Activity – Deeper thinking

1. How was smallpox wiped out? _By Vaccination world wide_

2. How long is one nanometre? _A Billionth of a metre_

3. What are the viral genes for? _To produce a virus inside a host cell_

4. How is the common flu virus transmitted from person to person? _By contact through Air, water and other Person_

5. How have big cities and air travel made humans more vulnerable to viruses? _High density means quicker spread of virus Air travel takes virus round the world in one or two days_

Army of viruses

How does a virus create an army inside a host? First it must enter a suitable cell. The cold virus prefers cells in the respiratory system. Some viruses inject their genes into a host cell. Others use their spikes to attach to a cell, then get absorbed into it. Once inside, the virus genes take command of the host cell.

The host cell is forced into making multiple copies of the virus. It's as if the virus uses the cell as a photocopier. Some cells make thousands of copies of the single invading virus. The virus copies then break out of the cell, and find new cells to repeat the process of infection. The whole infection process may destroy the host cells – contributing to what makes a person sick.

Bacteriophage

A virus that attacks a bacterium is called a bacteriophage.

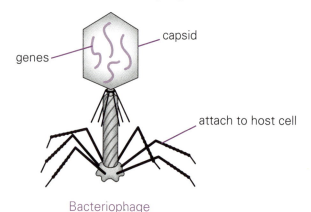

genes

capsid

attach to host cell

Bacteriophage

© ESA Publications (NZ) Ltd,
Customer freephone: 0800-372 266

Activity – Stages of attack

Write a sentence describing each stage of attack

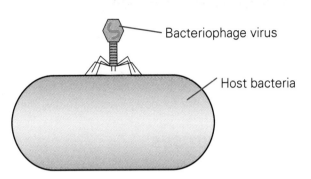

Bacteriophage virus

Host bacteria

1. _Virus ___ ___ bacteria_
   cell

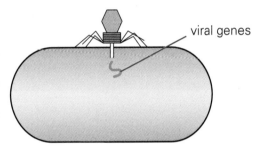

viral genes

2. _____

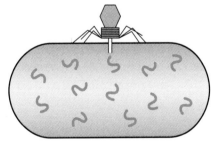

3. _____

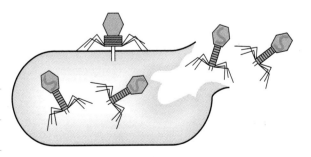

4. _____
   _____ host cell_

*AO: Explain how living things are suited to their particular habitat and how they
respond to environmental changes, both natural and human-induced.*

Science history: Vaccination

Edward Jenner made the first big breakthrough in the fight against viruses – the invention of vaccination. In 1796, a plague of the smallpox virus was killing thousands of people, but Jenner noticed that people who'd had a minor sickness, called cowpox, were not catching smallpox. He vaccinated people with cowpox fluid and they became immune to smallpox. He called it a *vaccine* after the Latin word for cow, 'vacca'. Today, vaccinations save many lives.

Bird flu viruses

Birds carry many different types of influenza virus. The H5N1 type of flu is especially lethal in chickens – a single gram of faeces can infect 40 million birds. Bird flu viruses are normally confined to birds, but in recent decades, H5N1 has infected some humans in contact with chickens. Viruses can mutate (change) quickly. The concern is that the H5N1 bird flu virus will mutate so it is able to jump from human to human.

Activity – Flu research

Find out when these influenza epidemics happened and approximately how many people died.

Influenza type	Years	Death toll (estimated)
Spanish flu		
Asian flu		
Hong Kong flu		

Group Activity

All viruses cause disease. Invite a doctor or scientist to class to discuss viruses. Find out how the effects of viruses can be reduced or prevented.

AO: Explain how living things are suited to their particular habitat and how they respond to environmental changes, both natural and human-induced.

© ESA Publications (NZ) Ltd,
Customer freephone: 0800-372 266

Life processes

Unit 6   Seeds

Flowering plants

Seeds are part of the reproductive cycle of flowering plants – which also includes pollination, fertilisation, fruiting, dispersal and germination.

Seeds form inside flowers. Flowers contain male and female sex cells. Pollen grains (produced by the anthers) contain male cells. The female sex cells (the ova) are inside the ovaries. To begin a seed, pollen must be transferred to the female part of the flower – this is called pollination.

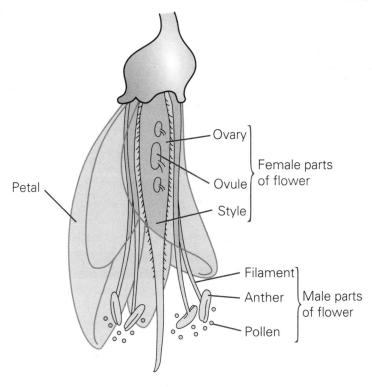

Kowhai flower

Flowers may use wind, insects (e.g. bees) and birds to transfer pollen. The pollen grain joins with an egg cell (ovum), which then develops into a seed – this is fertilisation.

As the seed grows, the flower parts die away, and a fruit develops around the seed. Fruits protect the seeds and help them in being spread away from the parent plant. A plant will try to disperse seeds as far as possible by wind, water, or animals.

A seed consists of an embryo plant, a food store, and a protective covering. When a seed lands in a place, it will usually only germinate when there is sufficient water and warmth. The embryonic plant starts to grow using its internal food store.

AO: Recognise that there are life processes common to all living things and that these occur in different ways.

Activity – Reproductive cycle

Label the stages in the reproduction of this flowering plant using the following words:

Fruiting   Pollination   Fertilisation   Germination   Dispersal

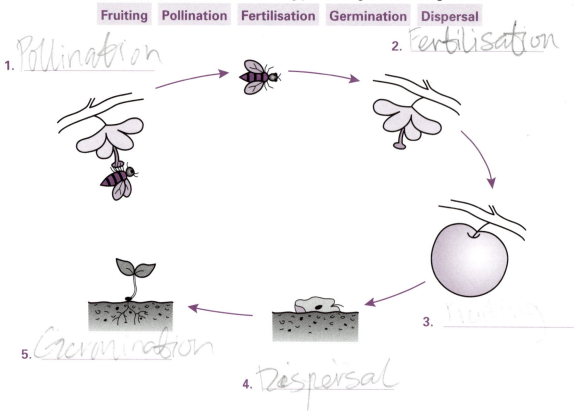

1. Pollination

2. Fertilisation

3. _____

4. Dispersal

5. Germination

Activity – Seed containers

Seeds are enclosed in a variety of 'containers'. Give an example of each of the following:

	Seed container	Example of a plant
1.	Pip fruit	
2.	Stone fruit	
3.	Berry	
4.	Pod	
5.	Gourd	
6.	Cob	
7.	Shell	
8.	Husk	

© ESA Publications (NZ) Ltd, Customer freephone: 0800-372 266

Seed dispersal

Plants reproduce with seeds, but seeds are also the way that plants spread out to new locations. If seeds aren't dispersed, they will have to compete with the parent plant. Dispersal depends on size – smaller seeds are usually spread by wind or explosion, heavy seeds are usually spread by animals.

New Zealand hook grass

- **Wind** – seeds must be light and small to be wind-carried. Seeds have developed adaptations to improve wind travel, such as fluffy hairs.

  Dandelions have parachutes that take seeds long distances.

  Sycamores have a wing on the fruit, which makes it spin away from the parent tree.

  Poppies have a pepper-pot container with openings that scatter seeds in the wind.

- **Water** – seeds from pond plants can float on water. Coconuts are a hollow waterproof shell. Coconuts have been found to be carried up to 1 600 km by the sea.

- **Explosion** – seeds can be ejected from exploding pods. Gorse and broom have dry seed pods which twist as they pop open, flicking out the seeds.

- **Animals** – animals eat fruit containing the seeds, but the seeds are not digested. Animal droppings that contain seeds provide ready-made plant fertiliser!

  Seeds may also have hooks and barbs which attach to animal fur (e.g. barley grass). Seeds such as acorns are collected by animals, but aren't always eaten.

Group Activity

Create a card or board game which teaches about seed dispersal. The game could include seed dispersal adaptations and means of dispersal.

© ESA Publications (NZ) Ltd,
Customer freephone: 0800-372 266

AO: Recognise that there are life processes common to all living things and that these occur in different ways.

Activity – Seed adaptations

Complete this table:

Species	Seeds dispersed by	Seed adaptation
Sycamore		
Scotch thistle		
Hook grass		
Coconut		
Pond iris		
Broom		

Activity – Growing seeds

Plants from warm countries will not always fruit in New Zealand because the summers are not long and hot enough. However, the following plants can be sprouted and grown inside, or outside in the warmer parts of New Zealand.

- Peanuts – soak raw peanuts overnight. Sprout on wet cotton wool.

- Dates – place in a pot of soil, seal the pot in a plastic bag, put in a warm place.

- Ornamental gourd (a plant from Central America) – follow instructions on seed packet.

- Avocado – sprout an avocado stone by suspending it in a jar of water in a warm place.

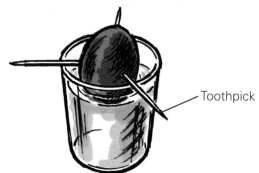
Toothpick

AO: Recognise that there are life processes common to all living things and that these occur in different ways.

© *ESA Publications (NZ) Ltd,*
Customer freephone: 0800-372 266

Unit 7   Digestive system

Breaking down food

Food is the raw material for cell growth – it gives muscles energy, and it keeps you healthy. The digestive system is adapted to extract nutrients from food. Adaptations are mechanical (teeth, stomach and intestine muscles) and chemical (enzymes). Food nutrients are mainly locked in large molecules that need to be broken down (digested) into small molecules.

Food travels about 9 metres and takes 24 hours to pass through the human digestive system.

- Firstly, food enters the **mouth**, where the teeth crush the food into smaller pieces. **Salivary glands** add an enzyme to start chemical breakdown.

- Next, the food is pushed down a tube called the **oesophagus** by muscles. The muscles are powerful enough to swallow even if you were standing on your head.

- The **stomach** partly digests the food with enzymes, and acid kills bacteria. Stomach muscles mechanically churn the food.

- Food then enters the small intestine, which is long and narrow. The first part is the **duodenum**, where chemical digestion is completed. The second part is the **ileum**, where food nutrients are absorbed into the bloodstream.

- Any undigested food enters the **large intestine**, where water is absorbed – leaving semi-solid waste. Finally, faeces collects in the **rectum**.

Activity – The journey of food

Label the parts of the digestive system using words highlighted above.

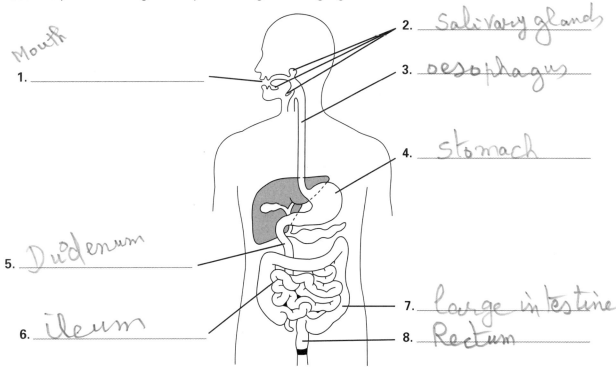

1. Mouth
2. Salivary glands
3. oesophagus
4. stomach
5. Duodenum
6. ileum
7. large intestine
8. Rectum

AO: Recognise that there are life processes common to all living things and that these occur in different ways.

Essential nutrients

A balanced diet provides the right amount of nutrients to keep healthy.

- The main energy nutrient is carbohydrate, found in bread and potatoes.

- Lipids (oils and fat) are an important energy source, but are only broken down after carbohydrate stores are used up. Lipids also contain a lot more energy than the same weight of carbohydrate. Foods high in lipids – such as deep-fried foods – should be eaten only occasionally.

- Protein is an essential nutrient for growth – it's found in fish, meat, eggs, beans, and dairy products.

- Vitamins and minerals are vital for health – fruit and vegetables are a good daily source.

Healthy food pyramid

The stripes on the pyramid represent the different food groups. Starting at the left-hand side, colour the triangle shapes as follows:

Orange   Green   Red   Yellow   Blue   Purple

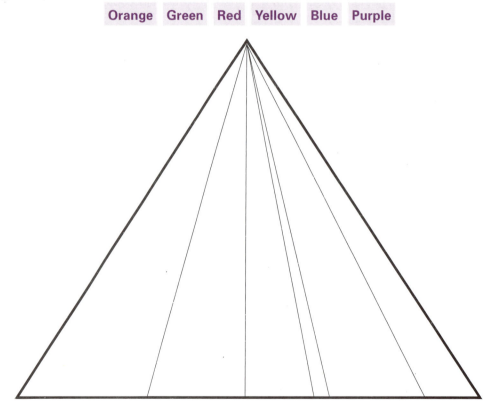

Orange = Grains (for example, bread, cereals, pasta, rice)

Green = Vegetables

Red   = Fruits

Yellow = Fats and oils

Blue   = Dairy products (for example, milk, cheese, yoghurt)

Purple = Meat, poultry, fish, beans, eggs, nuts

AO: Recognise that there are life processes common to all living things and that these occur in different ways.

© ESA Publications (NZ) Ltd, Customer freephone: 0800-372 266

Activity – Healthy food pyramid: True or False?

1. Processed grains are better to eat than whole grains.   True / False
2. Vegetables contain fibre.   True / False
3. Fruits are naturally low in fats.   True / False
4. You should avoid dairy products with low-fat content.   True / False
5. Beans and nuts are a source of protein.   True / False

Shocking science: Hole in the stomach

Dr William Beaumont discovered a lot about how digestion works in the 1820s. He saved a man who had a shotgun wound in his stomach. The wound left a hole through which Beaumont could examine the stomach. Beaumont tied food to silk threads and lowered it into the hole. Beaumont also collected stomach contents, placed them in glass tubes, and observed how long it took for the food to break down. He found the stomach used acids during digestion.

Group Activity

There are many scientific reports about food in the media, sometimes with confusing messages about what is a healthy diet. Discuss how we should respond to food advice. Come to a consensus about the generally agreed features of a healthy diet.

© ESA Publications (NZ) Ltd,
Customer freephone: 0800-372 266

AO: Recognise that there are life processes common to all living things and that these occur in different ways.

PLANET EARTH AND BEYOND
Astronomical systems

Unit 1   The Universe

Size

The Universe is everything that we know exists.
What appears to be mostly empty space
between the stars contains gases, radiation,
and 'dark energy'. The radiation includes
light rays, radio waves, and gamma rays.
Dark energy is a great scientific mystery –
it seems to act as a force against gravity,
making the Universe expand.

Gravity is an important force in holding the
Universe together. It keeps the planets in orbit,
and holds the stars in orbit in galaxies. The Sun is
one of 100 billion stars in the Milky Way galaxy. The
Universe has over 100 billion billion stars, grouped in
galaxies.

Distances in space are so vast they are measured in light years. A light year is 9 460 billion km – the
distance that light travels in space in a year. Our nearest galaxy – Andromeda – is 2 million light years
away. Compare this with the 8 minutes it takes light to travel from the Sun to Earth!

- Does the Universe have an end?

- Does the Universe go on for infinity?

- If the Universe does have an 'edge', then what lies beyond that boundary?

Astronomers can see objects about 12 billion light years away in the Universe. This is our 'visible
horizon'. No one knows what lies beyond this horizon, but it's thought that the visible Universe could
be a small part of something larger. There may even be other Universes that we don't know about.

Activity – Universe quiz: True or False?

Circle the correct answer.

1.	Space is mostly empty.	True / False
2.	Dark energy is not fully understood.	True / False
3.	Gravity makes the Universe expand.	True / False
4.	Sunlight is 8 minutes old by the time it gets here.	True / False
5.	The visible horizon is as far as astronomers can see at present.	True / False
6.	There is only one Universe that we know exists.	True / False

*AO: Investigate the components of the solar system, developing
an appreciation of the distances between them.*

© ESA Publications (NZ) Ltd,
Customer freephone: 0800-372 266

Science history: Einstein's ideas

One of Albert Einstein's greatest theories is that of General Relativity (1915), often called "the greatest leap of the scientific imagination in history". General Relativity states that the gravity of an object (e.g. a star) can distort time and space. It has been proved correct by observing the bending of starlight by gravity, and the behaviour of black holes. Einstein's ideas led directly to the Big Bang theory, which explains the origin of the Universe.

Expanding Universe

The Universe is about 14 billion years old. How did it begin? Most evidence points to a sudden Big Bang – a massive explosion and a vast expansion of size. We know that the Universe is still expanding, so it must have started off very small. Astronomers have detected the 'afterglow' of the heat from the Big Bang. Nobody knows what came before the Big Bang, or what created the 'seed' of the Universe.

At the beginning of time (t = 0) there was a tiny speck of concentrated energy – an incredibly hot fireball – which immediately started expanding. In the first billionth billionth billionth of the first second, the fireball 'blew up' to trillions of times its original size. The energy in that first second of time was so intense that subatomic particles of matter were created.

By the end of the first 3 minutes of time, the first elements – hydrogen, helium and lithium – were formed. Then the Universe settled down while it was cooling.

Then when the Universe was about 300 000 years old, light radiation began to escape and the Universe became clear, transparent space. The Universe was still expanding, but gravity now worked to pull the elements together. By 300 million years, stars and galaxies had formed in the Universe. The Universe continued, and continues, to expand.

The Milky Way galaxy formed at about 9 billion years, and about one billion years after that, the first life appeared on Earth.

Activity – The Universe time line

Write in some of the main events in the formation of the Universe:

Time	Event
0	Speck of energy
billionth billionth billionth of the first second	Fireball exploded
3 minutes	Hydrogen, helium, lithium formed
300 000 years	Universe became transparent
300 million years	Stars & Galaxies formed
9 billion years	Milky way formed
10 billion years	First life on earth
14 billion years	Present Day

© ESA Publications (NZ) Ltd,
Customer freephone: 0800-372 266

AO: Investigate the components of the solar system, developing an appreciation of the distances between them.

Black holes

No one has ever seen a black hole, but astronomers know they are out there. There's evidence that massive stars can turn into black holes. When a star explodes in a supernova, its core may collapse, crushing all matter into a small area of intense gravity. This gravity is powerful enough to suck in nearby stars. The gases that whirl around black holes get so hot that they release X-rays. Satellites have measured these X-rays. The first black hole discovered was Cygnus X-1, detected in 1970 by the satellite *Uhuru*. There may be a black hole at the centre of our galaxy – a ring of hot gas has been detected whirling around a strong gravitational centre!

The gravity of a black hole distorts nearby space-time. If you fell into a black hole, your body would be stretched like spaghetti, and time would run slower for you.

Activity – Deeper thinking

1. What has Einstein's theory about gravity got to do with black holes?

2. Why might time be changed in a black hole?

3. Why are satellites used for detecting X-rays from black holes?

Group Activity

Construct a large-scale timeline illustrating the development of the Universe since the Big Bang.

AO: Investigate the components of the solar system, developing an appreciation of the distances between them.

© ESA Publications (NZ) Ltd, Customer freephone: 0800-372 266

Unit 2   The planets

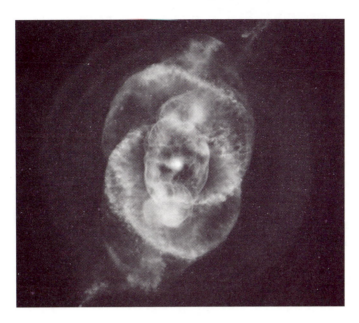

Cat's Eye nebula

Birth of the solar system

About 5 billion years ago, the Sun and the planets formed from a cloud of gas and dust called the solar nebula. The nebula collapsed and condensed into a disc of spinning gas – mostly hydrogen. The hot centre became the Sun. The material circling the Sun began to clump together to form protoplanets.

The protoplanets had enough gravity to attract more material, and some collided with each other. Smaller bodies were thrown off into space to become comets. By 3 billion years ago, the planets as we know them had settled into fixed orbits.

The planets closer to the Sun are made of rock and metal. During their formation, the Sun melted materials such as ice, and pulled in gases with its gravity. Far from the Sun, the outer planets are mostly made of gas and ice. During their formation, temperatures were lower, gravity weaker, and ice and gas survived more easily on these planets.

Planet Earth formed just the right distance from the Sun for life to develop.

* Temperatures aren't too hot or cold – both ice and liquid water can exist.
* The atmosphere traps breathable gases.
* There's a solid rocky surface to live on.

Debris from the original solar nebula is still scattered around the solar system – such as asteroids and ice dwarfs.

Group Activity

Create a class diagram or model of the birth of the solar system. Represent the formation of the solar system with different materials or with moving parts.

*AO: Investigate the components of the solar system, developing
an appreciation of the distances between them.*

Activity – Stages in formation of the solar system

Illustrate each stage in the formation of the solar system.

1. Solar nebula composed of gas and dust.

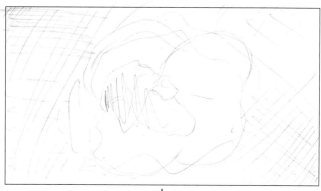

2. Nebula condenses into a spinning disc around the Sun.

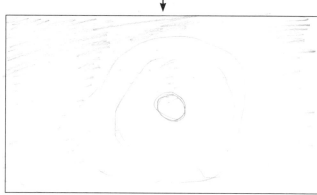

3. Protoplanets attract material and collide with each other.

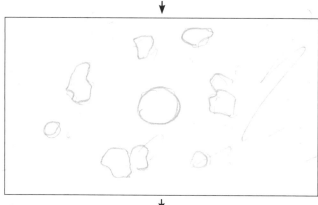

4. Planets settle into their present orbits.

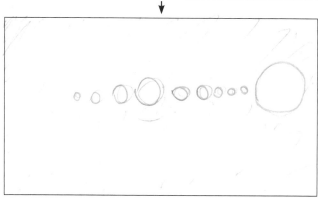

AO: Investigate the components of the solar system, developing an appreciation of the distances between them.

© *ESA Publications (NZ) Ltd,*
Customer freephone: 0800-372 266

Experiment: Spinning nebula

Material in space often forms into spinning discs around a dense centre. Examples include the condensing nebula, planetary rings, and black holes. You can see this effect in a small bowl (e.g. breakfast bowl) full of water. Add about a teaspoon of used tea leaves to the bowl and stir so they are whirling around. Allow the tea leaves to settle.

What do you observe about their movements? _____

The rock planets

The rock planets are made of solid rock with an iron core.

Mercury

Closest to the Sun, it has a wide temperature range (+450 °C to −180 °C) because it spins so slowly. It takes 58 Earth days to spin once on its own axis. It has hardly any atmosphere.

Venus

The surface is covered by thick, highly toxic clouds (mostly CO_2 and sulfuric acid). This atmosphere traps heat, making it the hottest planet, with an average temperature of 464 °C. The surface has a crust similar to Earth's crust, but with much larger volcanoes. It spins very slowly – it rotates once every 243 Earth days.

On Venus, a day is longer than a year! A year (time to orbit the Sun) passes in 224 Earth days, but a day (time to rotate once) takes 243 Earth days.

Activity – Finding Venus

Venus is the brightest planet in our sky. It's best seen in the east during the 3 hours before sunrise, and again in the west during the 3 hours after sunset. Never look at the Sun!

© ESA Publications (NZ) Ltd,
Customer freephone: 0800-372 266

AO: Investigate the components of the solar system, developing an appreciation of the distances between them.

Extreme science

Venus is not a comfortable planet to land on – the rain is acid, the air is poisonous, the pressure crushing, and the heat sizzling. Yet, in 1982, the Russian spacecraft *Venera* landed on the surface. It managed to send back the only colour photos of the surface – then the heavily armoured craft was destroyed by deadly Venus. Following is a photo of the surface of Venus.

Earth

Temperature range from –70 °C to 55 °C.

Mars

A cold planet which has no liquid water, but has ice caps. Temperatures range from –143 °C at the polar caps, up to 30 °C on rare summer days. The surface is a rusty-red colour, because of iron in the rocks and soil. Atmosphere is CO_2, with no breathable oxygen. It has the biggest volcano (25 km high) and the biggest canyon in the solar system. Day length same as Earth's day length.

The gas giants

The gas planets are made of gases, with no solid surface.

*AO: Investigate the components of the solar system, developing
an appreciation of the distances between them.*

Jupiter

The largest planet (142 984 km wide), made of hydrogen and helium. It spins very fast, taking only 9.93 hours to spin once – this makes the clouds that surround it form bands. The cloudy atmosphere has massive storms – its Great Red Spot is a hurricane nearly three times as wide as the diameter of Earth. Temperature averages –110 °C.

Saturn

The second largest planet has narrow rings made of chunks of ice and rock. The atmosphere has white ammonia clouds, and very strong winds. Average temperature is –140 °C. Spins once every 10.66 hours.

Uranus

A very cold planet (averages –197 °C) surrounded by thin rings. The atmosphere is a blue-green colour because of methane gas.

Neptune

Very cold (averages –200 °C) and dark. Has the most extreme weather of all the planets, with wind speeds up to 1 200 km/h. Atmosphere of hydrogen, helium and methane.

Activity – Planet table

Fill in this table describing conditions on the planets (most answers are in the text above).

Day length is the time it takes a planet to spin on it own axis.

	Temperature (°C)	Climate conditions	Atmospheric gases	What is the surface made of	Day length
Mercury					
Venus					
Mars					
Jupiter					
Saturn					
Uranus					
Neptune					

© ESA Publications (NZ) Ltd,
Customer freephone: 0800-372 266

AO: Investigate the components of the solar system, developing an appreciation of the distances between them.

Earth systems

Unit 3   Tectonics

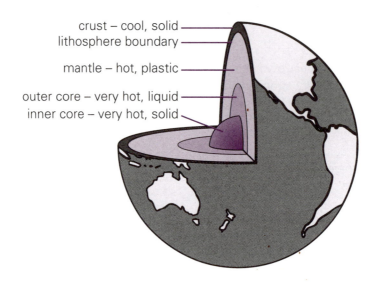

crust – cool, solid
lithosphere boundary
mantle – hot, plastic
outer core – very hot, liquid
inner core – very hot, solid

Layered planet

Planet Earth is made of separate layers – core, mantle and crust. The inner core is 1 200 km across, very hot (5 000 °C), and probably solid metal. The outer core is 2 250 km thick, 4 000 °C, and made of liquid metal. Around the core is the mantle, a 2 900 km thick layer of solid rock, with a temperature of 3 500 °C. The lower mantle rock is 'plastic' – it can flow very slowly. The upper part of the mantle is a layer of more rigid rock 100 km thick, and at a temperature of 1 000 °C. The outer layer is the cooler, rocky crust, which is up to 8 km thick under the oceans, and 70 km thick under the land.

Activity – Earth layers

Complete the data on this table:

Layer	Inner core	Outer core	Lower mantle	Upper mantle	Crust
Composition	Metal				
Solid or Liquid		Liquid			
Thickness (km)			2 900		
Temperature (°C)				1 000	

Tectonic plates

The rock inside the mantle is slowly moving in circular convection currents – heated rock rises to the surface, cools, sinks again, is heated again, and rises again. These heat currents create movement in the crust which 'floats' on top.

AO: Develop an understanding that water, air, rocks and soil and life forms make up our planet and recognise that these are also Earth's resources.

© ESA Publications (NZ) Ltd,
Customer freephone: 0800-372 266

The Earth's crust is broken up into a jigsaw of 17 massive rocky plates. The study of plate movements is called tectonics. Plates move slowly, at a speed of about 5 cm a year.

The boundaries (edges) where plates meet are places where there are volcanoes, earthquakes, mountain ranges and ocean trenches. There are three main kinds of plate movement – plates can slide past each other, plates can move towards each other, or plates can move apart.

- When plates slide past each other, friction between the rocks causes earthquakes. The San Andreas fault line in California is where the Pacific and North American plates grind past each other.

- When plates collide (move together), this can sometimes cause the land above to crumple into mountains. This is how the Himalayas were formed 50 million years ago when two continental plates collided.

  If one plate sinks down under another plate at the boundary, it's called subduction. When an oceanic plate 'dived' under the edge of the South American plate, the Andes mountain range was pushed up.

- When plates move apart, sea water may fill the gap. When the Arabian plate moved away from the African plate, the Red Sea was formed. You can see how the plates fitted on the map alongside.

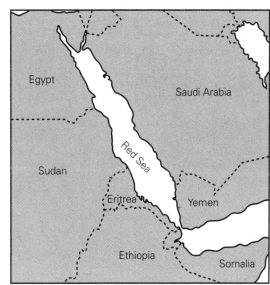

Activity – Tectonics

Fill in the gaps in the following paragraph:

The mantle is made of slow moving **1.** _____. Convection currents in the

mantle are caused by the rising of **2.** _____. The Earth's crust is fractured

into **3.** _____. Tectonics is the science of **4.** _____

_____. Earthquakes frequently occur at plate **5.** _____.

An example is the **6.** _____ _____ Fault. The Himalayas

were pushed up by the collision of **7.** _____ plates. The Andes are the result of

8. _____.

*AO: Develop an understanding that water, air, rocks and soil and life forms
make up our planet and recognise that these are also Earth's resources.*

New Zealand on the edge

The Pacific plate is the largest in the world. It's also the fastest moving (15 cm a year), because it's so thin. New Zealand sits on top of the collision boundary between the **Pacific plate** and the **Australian-Indian plate**. The collision boundary of the plates runs almost diagonally across the whole country. The forces of plate collision are shaping New Zealand by pushing, stretching, and sinking the land. It explains why we have so many mountains, earthquakes and volcanoes.

The whole country is being stretched as the plates move in opposite directions – Auckland is moving away from Christchurch at a rate of 1 metre every 20 years!

- East of the North Island, the Pacific plate is sinking under the Australian-Indian plate – it's called a **subduction zone**.

- There's another subduction zone in the Fiordland area, where the Australian plate is sinking under the Pacific plate.

- In the South Island, the two plates are colliding to push up the **Southern Alps**.

Activity – New Zealand plates

Label the diagram using the words highlighted above.

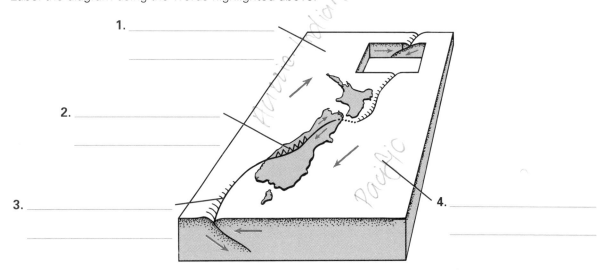

1. _____

2. _____

3. _____

4. _____

Group Activity

Look at several different websites that teach about tectonics. Evaluate the websites as a class. For example, you might compare the effectiveness of visual media, diagrams, models or explanations.

AO: Develop an understanding that water, air, rocks and soil and life forms make up our planet and recognise that these are also Earth's resources.

© ESA Publications (NZ) Ltd,
Customer freephone: 0800-372 266

Unit 4   Earthquakes

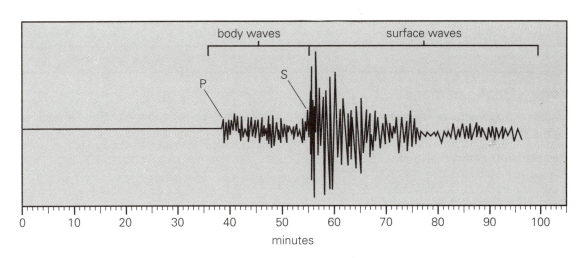

Seismic waves

Earthquakes usually happen along plate boundaries, where rocks are under enormous pressure. Older, more brittle rocks can fracture under pressure. The area of crust where the break happens is called a fault. A fault may sit for a long time as pressure builds up again. If the fault moves suddenly, it can cause an earthquake.

A quake starts at a focus point underground where the rocks break. This produces seismic waves, which are waves of energy which travel outwards. There are two main types of seismic waves – **body waves** and **surface waves**. Body waves can move deep under the Earth. Surface waves move near the surface of the Earth's crust, a bit like ripples.

Body waves come first from an earthquake and are called **P** and **S** waves. P waves (compressional waves) are felt first because they travel fastest. S waves (shear waves) arrive next. Surface waves arrive next, and are the waves that produce the most destruction on the land.

The **epicentre** of an earthquake is the place on the surface directly above the focus – it's where there's most damage.

The Richter scale

The Richter scale measures the amount of energy released by a quake. Each level of the scale is 10 times greater than the one before. For example, an earthquake measuring 6.0 on the Richter scale is 10 times stronger than a 5.0 quake. The largest earthquake recorded was a 9.5 in Chile in 1960.

Research the Richter scale and complete the table by describing the likely effects of an earthquake at each level of magnitude.

Magnitude on Richter scale	Effects and damage caused by earthquake
3.0–3.9	
4.0–4.9	
5.0–5.9	

© ESA Publications (NZ) Ltd,
Customer freephone: 0800-372 266

AO: Develop an understanding that water, air, rocks and soil and life forms make up our planet and recognise that these are also Earth's resources.

Magnitude on Richter scale	Effects and damage caused by earthquake
6.0–6.9	
7.0–7.9	
8.0–8.9	

Activity – Quake damage

Research the following earthquakes. Write a sentence for each describing some of the damage, and after-effects for people.

1.   Napier 1931 (7.8 on the Richter Scale): _____

_____

2.   San Francisco 1906 (8.3 on the Richter Scale): _____

_____

San Francisco 1906

3.   Boxing Day 2004, Sumatra-Andaman (9.0 on the Richter Scale):

_____

_____

4.   Haiti, 2010 (7.0 on Richter Scale) _____

_____

_____

5.   Japan, March 11, 2011 _____

_____

_____

AO: Develop an understanding that water, air, rocks and soil and life forms make up our planet and recognise that these are also Earth's resources.

© ESA Publications (NZ) Ltd, Customer freephone: 0800-372 266

South Island faults

The Alpine Fault in the South Island is New Zealand's largest fault (650 km long). It's also one of the longest natural straight lines on the planet – clearly visible from space. The plates here are jammed together under incredible strain. Major earthquakes happen on the fault about every 280 years.

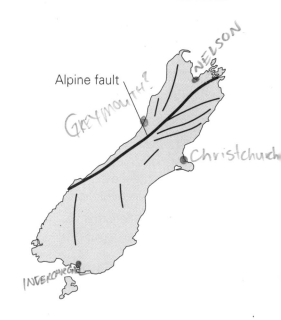

Christchurch earthquakes

In 2010, a magnitude 7.1 earthquake struck near Christchurch and caused widespread damage, but no loss of life. Thousands of aftershocks followed over the next few months. The quake was not caused by the Alpine Fault, but by a fault previously hidden. Then, on February 22, 2011, another quake struck Christchurch, measuring 6.3. This shallow quake was centred close to the city and released a huge amount of energy that caused severe shaking. Many buildings were destroyed, and 181 people died.

Christchurch quake

Group Activity

Research earthquakes in New Zealand over a month using a website such as geonet.org.nz

Plot the quakes on a map and discuss any patterns which emerge.

© ESA Publications (NZ) Ltd,
Customer freephone: 0800-372 266

AO: Develop an understanding that water, air, rocks and soil and life forms make up our planet and recognise that these are also Earth's resources.

Unit 5   Mountain building

Mountains are formed in the crust where tectonic plates collide. The two main processes that work together are folding and faulting.

- Folding occurs when rock layers bend under pressure.
- Faulting occurs when the pressure causes the rock layers to fracture.

New Zealand's Southern Alps were formed by folding and faulting. Firstly, the younger rocks folded upwards, then the older rock faulted into mountain-sized blocks. For 5 million years, the pressure of the plates has pushed the Alps up – and they are still rising by about 1 metre every 100 years.

Mount Everest (Sagarmatha, photo below) was formed as the Indian plate crumpled against the continental Eurasian plate. The 100 highest mountains (above sea level) are all in the Himalayas.

Activity – Mountain records

Research the heights and continents for these famous mountains:

Mountain	Continent	Height (m)
Everest		
K2		
Aconcagua		
Kilimanjaro		
Matterhorn		

Wearing down

Although mountains are being continually pushed up, they are constantly worn down again by two processes – weathering and erosion.

AO: Develop an understanding that water, air, rocks and soil and life forms make up our planet and recognise that these are also Earth's resources.

© *ESA Publications (NZ) Ltd, Customer freephone: 0800-372 266*

Weathering

Weathering is the wearing down of rocks by water, wind, and chemicals. Heating and cooling of rocks can make them crack. Water gets into cracks in a rock, then freezes and splits the rock apart. Chemical weathering can dissolve rocks – for example, acidic rainwater eats away at limestone. Plants and animals can also cause rocks to break down.

Erosion

Erosion is the movement of weathered rock and soil down slopes or in rivers. Erosion is carried out by wind, water, ice, and gravity.

- Wind blows smaller particles away, and rain washes material into rivers.
- Mountain rivers can carry large rocks. Rivers transport sediments to the sea.
- Gravity causes landslides and material falls down steep slopes. Mount Cook (Aoraki) was made 10 m shorter in 1991, when a chunk of rock and ice fell off the top, creating a massive landslide. Landslides can also be triggered by rainwater making soil heavier.
- Glaciers are solid blocks of ice that can cut through rock. The heavy ice grinds away rocks to form U-shaped valleys.

© ESA Publications (NZ) Ltd,
Customer freephone: 0800-372 266

AO: Develop an understanding that water, air, rocks and soil and life forms make up our planet and recognise that these are also Earth's resources.

Activity – Deeper thinking: Wearing down

1. How can heating and cooling crack a rock?

_____

_____

2. Why does ice split rocks?

_____

_____

3. What process formed the limestone caves at Waitomo?

acidic water ate away the limestone
to create caves.

4. What might happen if forests are cleared from hills?

_____

_____

5. Where in New Zealand would you find U-shaped valleys?

the south island

_____

Group Activity

Pairs or small groups research a different aspect of weathering or erosion, and present their findings to the whole class. Include visual media and demonstrations.

AO: Develop an understanding that water, air, rocks and soil and life forms make up our planet and recognise that these are also Earth's resources.

© ESA Publications (NZ) Ltd, Customer freephone: 0800-372 266

Interacting systems

Unit 6   Climate change

Ice ages and climate

The climate of planet Earth changes over long periods of time. It can be affected by changes in the Earth's orbit around the Sun, and changes in the Earth's tilt on its axis. Ice ages occur when the climate cools significantly – vast amounts of water are frozen into ice sheets.

An ice age means lower sea levels, cooler temperatures and much more severe winters.

Two million years ago, there was a major global ice age, which had dramatic effects on New Zealand. Temperatures dropped and massive glaciers formed. The Antarctic continent was iced over, which caused colder weather in New Zealand. The sea level dropped by up to 130 metres!

Since then, New Zealand has passed through about 20 cycles of ice ages. During these periods, the North and South islands were at times connected into a single land mass. This picture shows the extent of land above sea level about 20 000 years ago.

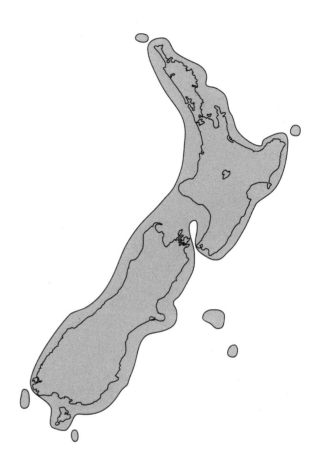

Activity – Thinking about ice ages

1. Why would the sea level drop during an ice age?

    _____

    _____

2. How would an ice age affect animals and plants?

    _____

    _____

© ESA Publications (NZ) Ltd,
Customer freephone: 0800-372 266

AO: Investigate the water cycle and its effects on climate, landforms, and life.

The greenhouse effect

Climate change due to ice ages is a natural process, but humans can have an 'unnatural' effect on climate. The temperature of planet Earth is slowly and steadily rising because of the 'greenhouse' effect (so-named because a greenhouse traps heat).

Earth is surrounded by an atmosphere composed of gases, including water vapour. When sunlight strikes the atmosphere, some of the solar energy is reflected back into space by clouds. The sunlight that strikes the Earth's surface changes to heat energy (infrared) that warms the planet. This heat energy is radiated by the Earth and escapes – but some of it is 'trapped' by gases in the atmosphere. Carbon dioxide, methane, and water vapour all absorb heat in the atmosphere. This heat is then re-radiated, and warms the Earth (and the atmosphere).

Activity – Global warming

Label the diagram using the following words:

Radiated   Absorbed   Reflected   Re-radiated

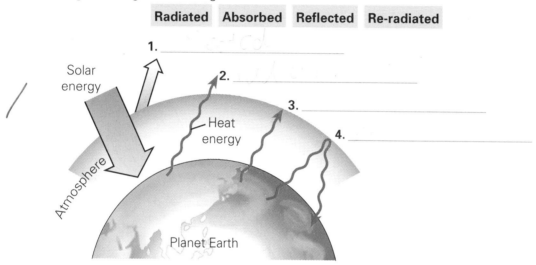

1. _____

Solar energy

2. _____

3. _____

Heat energy

4. _____

Atmosphere

Planet Earth

The human effect

Human activities have increased the amount of greenhouse gases in the atmosphere. This means there are more heat-absorbing gases. The CO_2 (carbon dioxide) level has increased mainly because of the use of coal and oil in industry and power generation (burning coal and oil releases carbon dioxide). Cutting down trees and burning forests also creates more CO_2. This is made even worse because trees actually remove CO_2 from the air as they grow. Car exhausts also add to CO_2. Methane is another greenhouse gas that humans add to, by farming cattle, from landfills and mining.

Group Activity

Discuss the global warming issue.

• How should we respond to predictions of climate disasters?

• What is our responsibility towards the environment?

Activity – CO_2 levels

Human activities have resulted in an increase in the amount of CO_2. In 1800, the level of CO_2 in the atmosphere was about 290 ppm (parts per million). In 1900, it was 295 ppm. By 1960, it was 315 ppm; in 1987 340 ppm; by 1999 360 ppm; 380 in 2005; and 388 ppm in 2010.

1. Complete this graph showing the rise in CO_2:

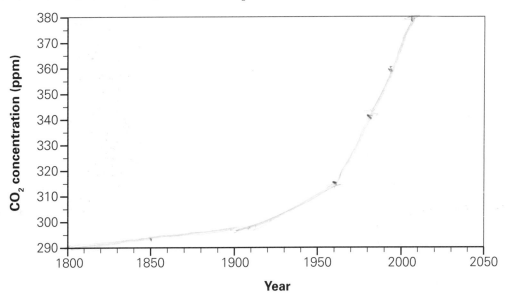

2. What do you notice about the rate of increase?

   Increases sharply starting 1960.

3. How might this relate to world population increase (3 billion in 1960 to 6 billion in 2000)?

   More people = more carbon dioxide.

Extreme science: Effects of global warming

A rise in surface temperature

This is already happening. Recent years have been among the warmest since modern records began to be kept in 1880.

- The decade 2000 to 2009 was the warmest on record.
- 2010 and 2005 were the warmest years, and tied for next warmest were the years 2009, 2007, 2006, 2003, and 2002.

Melting of polar ice caps and glaciers

In 2003, the huge Larsen B ice shelf broke off after 10 000 years of stability.

© ESA Publications (NZ) Ltd,
Customer freephone: 0800-372 266

AO: Investigate the water cycle and its effects on climate, landforms, and life.

Sea level rise and flooding of low-lying land

Sea levels have risen about 18 cm in the last 100 years. A rise of up to 43 cm, predicted for this century, would threaten major coastal cities.

Weather extremes such as bigger storms and droughts

The destructive Hurricane Katrina (photo) in 2005 was possibly made worse by warmer sea temperatures.

Activity – Sea level

1. Name three large New Zealand cities which are on the coast at sea level.

2. Name three large international cities at sea level on the coast.

3. What effect would a 43-cm rise in sea level have on a large city at sea level?

Reduce your carbon footprint

A carbon footprint is a way of picturing our impact on global warming. It is a measure of how much CO_2 a person adds to greenhouse gases. A person can reduce the size of their carbon footprint by reducing their CO_2 production in the following ways.

- Use cars less – walk, cycle, use public transport.
- Use less hot water – shower rather than have a bath.
- Turn off electric devices when they are not needed.
- Plant a tree.
- Use energy-efficient light bulbs.
- Recycle.

© ESA Publications (NZ) Ltd,
Customer freephone: 0800-372 266

Unit 7   Climate

Climate factors

Climate is the pattern of weather in a place over a long period of time. Temperature is the main factor in global climate; temperature changes with latitude. The Sun creates different climates as it strikes the curved surface of planet Earth at different angles.

- The equator gets more heating, because the Sun's rays strike the Earth almost directly (i.e. close to right angles or 90°).

- It's cooler towards the poles, because the Sun's rays hit planet Earth at a smaller angle – they cover a larger area and so have less heating power.

Equator

Climate is also determined by height above sea level. Temperature decreases with height, because 'thinner' air cannot hold the heat so well. It's possible to find snow and ice at the equator, such as on Mt Kilimanjaro in eastern Africa.

© ESA Publications (NZ) Ltd,
Customer freephone: 0800-372 266

AO: Investigate the water cycle and its effects on climate, landforms, and life.

Another climate factor is distance from the sea. The land heats and cools faster than the sea. Inland areas have hotter summers and colder winters. This is called continental climate. For example, Central Otago has a greater temperature range than coastal Otago.

Most of New Zealand has a temperate climate, with mild winters and warm summers.

The main climate zones are polar, tropical and temperate. Polar regions are very cold and dry all year, while the tropics remain hot and wet for much of the year. Temperate zones are in between, where the weather is changeable during the course of a year.

Activity – Climate terminology

Match the definitions with the words that follow:

Latitude   Continental   Temperate   Sea level   Tropics   Equator   Climate

Definitions:

1. Climate that changes during the year. _Temperate_

2. On a map, the position north or south from the equator. _Latitude_

3. A more extreme climate in inland areas. _Continental_

4. Average weather, or regular variations in weather in a region over a period of years.
climate

5. Measure from surface of the ocean and above. _Sea level_

6. Imaginary circle around Earth the same distance from the North and South Poles.
Equator

7. Places that remain hot and wet year-round. _Tropics_

Climate zones

The three main zones (polar, tropical, temperate) can be subdivided into the following nine smaller zones.

- **Polar/Tundra**. Polar areas have permanent ice. Temperatures in the tundra (around the Arctic circle) can rise above freezing level, so small plants grow there.

- **Boreal forest** – has long cold winters, and short cool summers.

- **Mountains** – have colder temperatures higher up. Usually have more rain, snow and wind than lowlands.

- **Temperate forests** – have distinct seasons, with mild winters and mild summers. Rainfall spread throughout the year.

- **Mediterranean** – cool winters and extremely dry summers.

- **Desert** – hot and very dry all year.

- **Dry grassland** – hot summers, cold winters. A little rain. Found in the centre of continents.

- **Tropical grassland** – hot all year. Has a wet season, then a dry season.

- **Tropical rainforest** – hot and humid, rainfall throughout the year.

AO: Investigate the water cycle and its effects on climate, landforms, and life.

© ESA Publications (NZ) Ltd, Customer freephone: 0800-372 266

Activity – Climate zones examples

Locate each of the following seven places on the world map. Alongside each place, identify its climate zone.

1. Himalayas: *mountain*

2. Northern Canada: *Tundra*

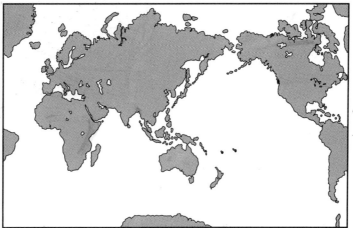

3. Italy: *Mediteranian*

4. Amazon Basin: *Tropical Rainforest*

5. African Savannah: *Tropical Grassland*

6. Central Australia: *Desert*

7. West Coast New Zealand South Island: *Temerate Rainforest*

Group Activity

Divide the class into nine groups. Each group represents one of the main climate zones listed on the previous page. Each group presents their zone as a chart, poster, or computer presentation.

AO: Investigate the water cycle and its effects on climate, landforms, and life.

Unit 8   Meteorology

Moving air

Accurate forecasting of the weather can save
lives, especially when storms are coming.
Meteorologists measure atmospheric conditions
including air pressure, temperature, wind, and
rainfall. The data is used to predict the weather.

Air pressure is the weight of the atmosphere.
Air is constantly moving around the planet, but it
tends to accumulate in some areas – this is high
pressure (H). Air always flows from a high-pressure area to a low-pressure area (L) – this flow makes
the wind.

- On a weather map, air pressure is shown by lines called isobars. Isobar lines connect points of
  equal pressure. The closer the spacing between the isobars, the stronger the wind.

- High-pressure systems (also called anticyclones) usually bring lighter winds and dry weather.
  The air in a high-pressure system sinks, which warms the air and evaporates clouds. Wind flows
  anticlockwise around a high-pressure area in the southern hemisphere.

- Low-pressure systems (called depressions) usually bring strong winds and wet weather. The air
  in low-pressure systems rises and cools, condensing into clouds and rain. Wind flows clockwise
  around a low-pressure system in the southern hemisphere.

- A front is the boundary where two masses of air meet. Fronts can create a sudden change in
  weather. A cold front is a mass of cold air pushing into an area of warm air. On a weather map,
  a cold front is marked with triangles; a warm front with semicircles.

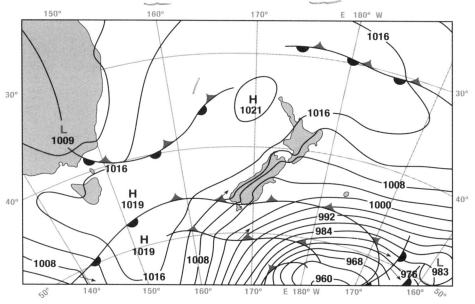

Typical spring weather in New Zealand – cold southerly front
bringing strong south-westerly winds

Weather map symbols

Explain these symbols from the weather map.

1. *High*

2. *Low*

3. *Isobars – show air pres...*

4. *Cold Front*

5. *Warm Front*

Activity – Technology: Make a barometer

A barometer measures atmospheric pressure. To make a simple barometer, you need a glass jar, a balloon, and a drinking straw. Cut a piece of balloon to fit the top of the jar. Stretch it over the jar and secure with a rubber band. Tape the end of a straw to the top. Place the jar alongside a piece of card, so the straw is almost touching it.

During the day, the straw will move slightly. Mark the position of the straw onto the card.

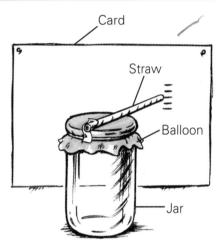

How it works

When the air pressure is higher, the rubber is pushed down – the straw rises. When air pressure drops, the rubber bulges up.

Observe changes for a few days.

* Does the change in air pressure match the weather outside?

* Is high pressure associated with settled weather?

AO: Investigate the water cycle and its effects on climate, landforms, and life.

Science history: Celsius

Anders Celsius (1701–1744) was an amazing astronomer and physicist who invented the Celsius thermometer scale for measuring temperature. He also discovered the cause of auroras, helped to prove the Earth was round, and built Sweden's first observatory.

Group Activity

Divide the class into small groups. Each group has to devise a way to measure one aspect of the weather – rainfall, wind, temperature, pressure, humidity, etc.

New Zealand weather

New Zealand weather is dominated by high- and low-pressure systems which move from west to east. A high brings settled weather, while a low brings unsettled conditions. The prevailing (most common) wind over much of New Zealand is from the westerly direction.

When the wind comes from the south it is colder because it comes from the cool Southern Ocean. A southerly front can be a very sudden, cold change that brings snow.

The mountains in New Zealand have an effect on weather.

- Air comes off the Tasman Sea and hits the Southern Alps. The air is forced up and it cools, dropping rain as it does so on the West Coast of the South Island. On the other side of the Alps, the air sinks and warms. This can create a warm northwest wind on the South Island east coast.

- A gap in the mountain ranges allows stronger winds to rush through. This happens in Cook Strait and Wellington.

Activity – Read a weather map

Write the correct letter in the box on the map that corresponds to the following.

A   A cold front moving north.

B   A strong northwest wind flow over inland Canterbury.

C   A high over the North Island.

D   Strong southerly winds.

E   A warm front.

F   Heavy rain in the west of the South Island.

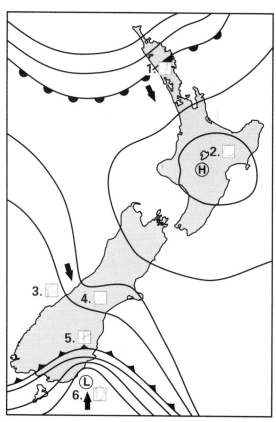

AO: Investigate the water cycle and its effects on climate, landforms, and life.

Physical inquiry and physics concepts

Unit 1   Motion

Dynamics

Dynamics is the science of force and motion. It's about the way an object moves when acted on by a force. Everything in the Universe moves, from the extremely small electron to massive galactic clusters.

Objects only change speed or direction because they are 'forced' to. Once moving, objects have their own momentum, which keeps them going at the same rate, and in the same direction, until a force opposes (changes) the movement. An object would keep moving in outer space where there's very little friction. On Earth, moving objects are slowed and stopped by forces of friction. (Gravity also pulls objects towards the Earth.)

A force that produces motion is usually a push or a pull. A force acts in one direction – but there's always a force acting in the opposite direction.

Activity – Movements

Choose words from those following to complete the statements below:

gravity   orbits   air   movement   directions

1. Dynamics is the study of _movement (force and motion)_ .

2. Satellites and planets move in _orbits_ .

3. A rocket in outer space is not slowed down by _air_ .

4. A falling ball is pulled towards the ground by _Gravity_ .

5. Forces act in opposing _directions_ .

Group Activity

Divide the class into small groups. Each group has to devise a way to demonstrate or prove one of the following concepts – momentum, acceleration, Newton's Laws of Motion.

Evaluate each group.

© ESA Publications (NZ) Ltd,
Customer freephone: 0800-372 266

Explore, describe, and represent patterns and trends for everyday examples of physical phenomena such as movement, forces, electricity and magnetism, light, sound, waves, and heat.

Science history: The first scientist?

Galileo Galilei (1564–1642) was one of the first scientists to use experimentation and mathematics to help to prove or disprove his ideas. He recognised that forces set objects in motion, and forces are needed to halt them. Galileo's theories about dynamics were later used by Newton and Einstein.

Activity – Galileo's legendary experiment

One of Galileo's ideas was that falling objects should be pulled towards Earth with equal accelerations. To test this, he is said to have dropped weights of differing masses from the top of the Leaning Tower of Pisa (photo below).

What would this experiment have shown about falling objects?

They would have fallen and hit the ground at the same time.

large and small masses fall at the same rate. Gravity pulls all masses masses with equal Force. Two different Sizes or weights will reach the ground at the same time as long as they are of the same same size shape.

Speed and acceleration

Speed is a measure of the distance moved by an object in a certain time – e.g. kilometres per hour (km/h) is the speed unit for a car. Most objects don't stay at the same speed. The fastest speed that we know of is the speed of light.

When an object increases its speed, this is called acceleration. Acceleration is caused by unbalanced forces. For example, when you are travelling only slowly on a flat section of road on a bike and then pedal harder, you accelerate – the pedal force is greater than the force of friction (air and road).

Acceleration is a measure of the gain in speed over a certain time period. For example, imagine a bicycle is moving at a speed of 1 metre per second (1 m/s). It accelerates to 2 m/s after a second passes, then to 3 m/s after another second passes. Its rate of acceleration is 1 m/s every second – written as 1 m/s^2.

The lighter the object and the larger the force, the greater the acceleration.

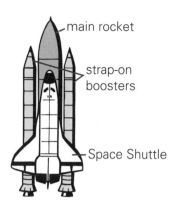

main rocket

strap-on boosters

Space Shuttle

Light object – very small force needed to accelerate.

Massive object – huge force needed to accelerate (booster rockets are used to accelerate shuttle away from Earth's gravitational pull).

Activity – Speed units

Match these moving objects with the correct number:

1.	A snail moves at	30 000 km/h
2.	A person walks briskly at	1.5 m/s
3.	A honey bee flies at	1 224 km/h
4.	Olympic sprinter runs at	1 000 km/h
5.	A racing car travels at	300 000 km/s
6.	A passenger jet flies at	10 m/s
7.	The speed of sound is	3 mm/s
8.	The space shuttle moves at	300 km/h
9.	The speed of light is	25 km/h

Collisions

Momentum helps us to understand what happens when objects collide with each other. Momentum is concerned with the movement of an object – an object will keep moving at the same speed until something slows it.

Explore, describe, and represent patterns and trends for everyday examples of physical phenomena such as movement, forces, electricity and magnetism, light, sound, waves, and heat.

Momentum = mass × speed

Momentum can be dangerous. Imagine sitting in the front seat of a moving car. The car has momentum, and so do you. The car brakes suddenly and stops – but you still have your forward momentum; if you have no seatbelt on, you will hit the windscreen.

Momentum depends on two things – the speed of an object and its mass. The heavier the car and the faster it goes, the greater its momentum – and the harder it is to stop!

Momentum can be passed from one object to another. For example, when a moving snooker ball collides with a stationary snooker ball, the stationary ball can gain the momentum that the moving ball loses.

Changes in momentum are always in the direction of an applied force. If the forces are internal, as in the case of the explosion that drives a bullet from a gun, then the momentum of the bullet is opposite and equal to the momentum of the recoiling gun.

Activity – Explaining momentum

1. Explain how momentum could injure a person standing on a bus that stopped suddenly.

    The person still has forward momentum so they will be thrown forwards

2. Explain how the momentum of a heavy demolition ball can knock down a brick wall.

    The momentum goes from the wrecking ball to the bricks, which makes the bricks move

3. Explain the effect of momentum in the photo of a meteorite crater.

    that heaps of momentum that made the crater

Unit 2   Gravity

The long distance force

Gravity is a force of attraction between any two objects in the Universe. The force of gravity is easily seen when one of the bodies is very large and the other is smaller. The Sun's gravity attracts the planets into their orbits. Planet Earth has a gravitational force field which makes us fall downwards, but stops us floating all over the place.

The force of gravity depends on the mass of a planetary body. The mass of the Moon is much less than the mass of the Earth – the Moon's gravity is about 17% of that of the Earth. The Earth holds the Moon in orbit, but the Moon also has a pull on the Earth (causing tides).

The force of gravity depends on distance. The Earth's gravitational pull becomes weaker as you move away from the surface.

Gravity pulls us towards the centre of the Earth. Gravity pulls all objects with a single force. If you drop two objects the same shape but different masses, they will hit the ground at the same time. Why? Because, though the more massive object experiences a greater gravity force, it is also harder for the gravity force to move it (because it is more massive). Therefore, heavy and light objects will fall at the same rate. Note that air resistance will make some objects fall more slowly.

Group Activity

In pairs, students think of questions about gravity and related concepts – weight and force. Share the questions with the whole class, then select five questions to ask a scientist or submit the questions online at www.anyquestions.co.nz or through newspapers.

Explore, describe, and represent patterns and trends for everyday examples of physical phenomena such as movement, forces, electricity and magnetism, light, sound, waves; and heat.

Activity – Gravity

Draw arrows to show where gravity is
acting between the astronomical bodies
shown – the Sun, Earth and Moon.

Note: draw arrows pointing in the
direction of attraction.

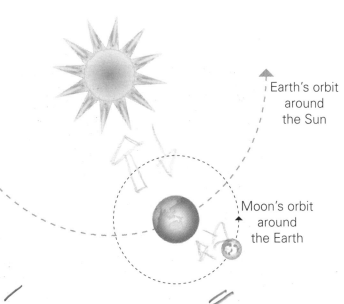

Earth's orbit
around
the Sun

Moon's orbit
around
the Earth

Weight is a force

The weight of an object is different
from the mass of an object. Weight is a
force – caused by the downward pull of
gravity. Your weight changes if gravity
changes. For example, on the Moon
you would weigh one-sixth of your
present weight.

Mass is the amount of matter in an object. Your mass is unaffected by gravity. Your mass would
be the same on the Moon as it is on Earth. Mass is measured in grams, kilograms, etc. Weight is
measured in newtons (N). To find your weight in newtons, simply multiply your mass in kg by 10.

Example:

$W = m \times g$

American Neil Armstrong became the first man to walk on the Moon on 21 July 1969.

On Earth, Neil Armstrong had a mass of 80 kg and a weight of 800 N ($80 \times 10 = 800$).

On the Moon, Neil Armstrong had a mass of 80 kg and a weight of 133 N ($80 \times 10 \times 1/6 = 133$).

Activity – Explaining mass and weight

1. How high could you jump on the Moon? _Six times high than on earth_
    higher than on earth

    Why? _Moons gravity is 1/6 less than on Earth_

2. a. What is your mass on the Moon? _Same as on earth_

    b. What is your weight on the Moon? _one sixth ($\frac{1}{6}$) as on earth_

3. Why will a sheet of paper float slowly to the ground instead of falling straight down?

    Because air resistance

    Air

4. What is your weight, in newtons? _400 N_
    ten times your mass

Unit 3   Floating and sinking

Why objects float

Gravity acts on all objects, even in water. But floating objects also experience a force called upthrust. When an object is placed in water, its weight pushes downwards – this is the force of gravity. At the same time, the water pushes back upwards (upthrust).

When placed in water, objects will push some water aside. Why do some objects float, while others sink? It depends on how much water the object pushes out of the way (displaces). Floating happens when the displaced water creates an upthrust large enough to support the object's weight.

Density is the most important factor in flotation. Density is the mass of a substance (of a given volume). For example, a golf ball is more dense than a table-tennis ball of the same size. Any solid substance denser than water will sink when placed in water. Metal and stone are denser than water – they will sink (unless made into a boat shape).

The shape of an object also influences flotation. Boats float well because they displace sufficient water. If an object can be shaped to displace enough water, it will float.

Activity – Floating and sinking definitions

Define these words:

1. Upthrust: *Force that Pushes upwards against object in water*

2. Displacement: *Water pushed aside by an object*

3. Floating: *displaced water creates enough up thrust to support an object*

4. Sinking: *Not enough up thrust to support an object*

5. Density: *Mass of a substance (of a given volume)*

© ESA Publications (NZ) Ltd,
Customer freephone: 0800-372 266

Explore, describe, and represent patterns and trends for everyday examples of physical phenomena such as movement, forces, electricity and magnetism, light, sound, waves, and heat.

Investigate: Floating

1. **Upthrust**

   Push a flat block of wood into water.

   What force do you feel? _____

2. **Displacement**

   Float three balls of different masses in a bowl of water (e.g. table tennis ball, tennis ball, plastic ball). Observe the amount each sinks (the amount of displacement).

   What do you notice about ball density and displacement of water? _____

   _____

3. **Shape**

   Drop a plasticine or clay ball into water. Now shape it into different boat shapes to make it float.

   Which shape floats best? Why? _____

Science history: Archimedes (287–212 BC)

Archimedes' Principle explained the force of upthrust. Legend has it he used the idea to test the king's crown to see if it was real gold. He found that the crown displaced more water than a gold bar of the same weight. Because the crown wasn't pure gold, the goldsmith was executed!

Group Activity

Students work in groups to make a toy boat that will carry a 2 kg weight. Brainstorm rules for a floating competition, including criteria for the construction, materials, and testing.

Explore, describe, and represent patterns and trends for everyday examples of physical phenomena such as movement, forces, electricity and magnetism, light, sound, waves, and heat.

© *ESA Publications (NZ) Ltd,*
Customer freephone: 0800-372 266

Buoyancy

Buoyancy is another word for upthrust – the upward force on an object in a liquid.

Objects seem to weigh less in water. For example, it is easier to lift a person while they are floating in water than to lift them when they are out of the water. In fact, objects do weigh less in water, because weight depends on all the forces acting, and buoyancy acts to reduce or cancel the effect of gravity. (The object's mass stays the same.)

Maintaining buoyancy is important for boats, human swimmers, and animals.

Activity – Research

1. Explain how a fish uses a swim-bladder to change its buoyancy.

2. Explain how a submarine can change its buoyancy so that it can float or sink.

3. Draw a scuba diver to show the equipment used to change a diver's buoyancy.

© ESA Publications (NZ) Ltd,
Customer freephone: 0800-372 266

Explore, describe, and represent patterns and trends for everyday examples of physical phenomena such as movement, forces, electricity and magnetism, light, sound, waves, and heat.

Unit 4   Energy

Forms of energy

Energy is not a material substance, yet it is essential for life. There are many different kinds of energy, including light, heat, chemical, electrical, sound and movement. Energy is needed to make things move, change, or heat up – this is called 'doing work'. The Sun is the original source of most of the energy on Earth.

Two basic types of energy are kinetic energy and potential energy.

- Kinetic energy is the energy of a moving object (from the Greek '*kine*', meaning movement). Kinetic energy depends on the mass and speed of an object – a fast-moving train has more kinetic energy than a bus.

- Potential energy is stored energy – it's not yet active. Potential energy can be chemical, nuclear, elastic, magnetic or gravitational. For example, food chemical energy is only released when it is eaten and used by the body. Objects above ground level are a store of potential energy – it's changed into kinetic energy when an object falls.

Activity – Kinetic or potential?

For each example, tick the box to show whether the energy involved is mainly kinetic energy or mainly potential energy.

Example	Kinetic energy	Potential energy
Bird flying fast near the ground	✓	
Heat from a fire	✓	
Skydiver in a plane		✓
Lightning flash	✓	
Petrol in a tank		✓
Bird flying high and slowly	✗	✓
Rolling soccer ball	✓	
Pizza		✓
Stretched rubber band		✓

Energy changes

Energy can be transferred from one object to another

When energy is transferred from one object to another, this process is called "doing work". Here are some examples.

- When you hammer in a nail, energy is transferred from your body to the hammer and into the nail, driving the nail into the wood.

- When you lift a load using a wheelbarrow, your body transfers energy into the wheelbarrow and the load in the wheelbarrow.

- When you kick a rugby ball, the kinetic energy from your moving foot is transferred to the ball.

- When you sit in the hot sun, energy is being transferred – heat energy of the Sun is transferred to your body and you get warmer.

Energy can be transformed into other forms

A skydiver stands in a plane. She has some potential energy of gravity. When she falls, this energy is transformed into kinetic energy.

An electric blanket (when it's switched on) is transforming electrical energy into heat energy.

Activity – Energy transfer

Write the sequence of energy changes that occur under the picture involving hitting a ball that smashes a window.

K.E

Produce Sound and heat energy from Broken window

Energy in = Energy out

Energy cannot be created or destroyed. Energy can only be changed from one form to another. This is called the Conservation of Energy. It may seem as if you can create energy, but it must always have come from somewhere. For example, you get energy from firewood, but the plant that made the wood originally got its energy from the light of the Sun.

Explore, describe, and represent patterns and trends for everyday examples of physical phenomena such as movement, forces, electricity and magnetism, light, sound, waves, and heat.

When energy changes occur, the total amount of energy always stays the same (Energy in = Energy out). Consider the electricity that makes a light bulb glow. Some of the electrical energy is changed to light energy, and the rest becomes heat – but the amount of energy that went into the bulb is the same as the amount that comes out.

Heat is produced in many energy changes. When the heat is not used for any purpose, the heat is called 'waste' or 'lost' energy. For example, a car engine produces a lot of heat which is not useful (apart from a little used in a car heater). An energy-efficient product is one that produces little waste energy. For example, energy-efficient light bulbs produce only 30% of the heat of normal bulbs.

Activity – Energy chain

Use the words following in the correct order to label the energy chain:

heat   kinetic   heat   chemical   light

1. _Heat_

    and

    Light

2. _Chemical_

    energy

3. _Kinetic_

    energy and

    Heat

*Explore, describe, and represent patterns and trends for
everyday examples of physical phenomena such as movement,
forces, electricity nd magnetism, light, sound, waves, and heat.*

© *ESA Publications (NZ) Ltd,
Customer freephone: 0800-372 266*

Unit 5   Renewable energy

Over 80% of the energy we use on Earth comes from burning fossil fuels (coal, oil and natural gas). These are non-renewable resources – they took millions of years to form. We are using them at an increasing rate, and the known reserves will probably run out during your lifetime. We need to look for energy sources that are more sustainable – sources that are constantly renewed by nature.

Sources of renewable energy

Sources of renewable energy include the Sun, water, geothermal, wind, biofuels and oceans.

Solar power

Sunlight is our greatest source of heat and light on Earth. It directly heats and lights our buildings, and it can be used to generate electricity and heat water.

- **Solar water heaters** have a collector and a storage tank. The collector is often on the roof of a building facing the Sun. The Sun heats water that passes through tubes inside the collector.

- **Solar cells** (photovoltaic devices) can change light directly into electricity. They are made of semiconductors which produce a flow of electrons (electricity) when they absorb sunlight. Solar cells can be small or large, powering calculators or whole buildings.

- **Concentrating solar power plants** use mirrors to concentrate the Sun's energy. This heats fluids which are used to run generators which make electricity.

© ESA Publications (NZ) Ltd,
Customer freephone: 0800-372 266

*Explore, describe, and represent patterns and trends for
everyday examples of physical phenomena such as movement,
forces, electricity and magnetism, light, sound, waves, and heat.*

Many solar cells are needed to power buildings

Activity – Solar limitations

List some of the limitations or drawbacks of solar power:

Hydroelectric power

Most hydroelectric power in New Zealand is made from water stored behind a dam on a river. The higher the level of the stored water, the more energy can be converted into electricity. As the water is released, it spins a turbine that generates electricity.

Group Activity – New Zealand hydro power

Hydroelectric power stations are the largest source of electricity in New Zealand. It's a renewable source, but there are often protests against building dams because of damage to the natural environment. Debate the issues involved in building hydro plants. Consider the alternatives available, economic benefits and environmental impacts.

Geothermal power

Heat is produced deep within the Earth by the decay of radioactive particles in rocks. Geothermal energy is from areas where the heat finds an outlet or where the Earth's crust is thin. Heat comes to the surface in volcanic areas, in hot springs and geysers. Geothermal power stations tap into reservoirs of hot water underground and use the steam to drive turbines that generate electricity. New Zealand has large-scale geothermal power stations around the Taupo volcanic zone – they

© ESA Publications (NZ) Ltd,
Customer freephone: 0800-372 266

provide about 10% of the country's electricity. The Ngawha geothermal field provides up to 75% of the Far North's energy. Buildings can also be heated directly from hot water reservoirs near the surface.

Steam pipes at Wairakei Power Station

Activity – Other sources of renewable energy

Investigate other sources of renewable energy – such as wind power, biofuels, ocean power and hydrogen power. Create a chart that lists the advantages and disadvantages of using each source of energy to generate electricity.

Nuclear power

Nuclear energy is sometimes called renewable; however, it uses a source of fuel that could run out – it is limited just as oil and coal are. One advantage is that nuclear power does not produce greenhouse gases.

Science history: Nuclear melt-down

In 1986, a nuclear power reactor exploded at Chernobyl in the Ukraine. It released 100 times more radiation into the air than the atom bomb dropped on Hiroshima in World War II. Radioactive deposits from the Chernobyl explosion spread by wind and rain over the whole northern hemisphere. There is still a 1 000 sq km exclusion zone of contaminated land around the reactor. 350 000 people were forced to move, and many cancers were caused.

Find out what happened at the **Fukushima Daiichi Nuclear Plant in 2011**.

© ESA Publications (NZ) Ltd,
Customer freephone: 0800-372 266

*Explore, describe, and represent patterns and trends for
everyday examples of physical phenomena such as movement,
forces, electricity and magnetism, light, sound, waves, and heat.*

Unit 6   Heat

Conduction Solid
Convection liquid
Radiation

Conductors and insulators

Heat is the kinetic energy of particles (atoms and molecules). An increase in the temperature of a substance means an increase in the kinetic energy of its particles. When particles are heated they move faster.

- Particles in a solid vibrate faster as they heat up.
- Particles in a liquid and a gas move faster and further as they heat up.

When heat moves through solid objects, it's called conduction. Vibrating particles in a solid pass their energy along the object, from particle to particle. Some materials conduct heat more effectively than others. Most metals – including iron, silver and aluminium – are good conductors. This is because the atoms in metals are closely packed, so vibrations are quickly passed on to neighbouring atoms. Metals have another advantage too – they have a lot of 'free electrons' drifting through them. The free electrons help to pass heat more quickly.

A poor conductor of heat energy is called an insulator. An insulator is a good material for keeping you warm. Air is one of the most effective insulating substances – particles in air are far apart and the collisions needed to transfer heat occur infrequently compared with those in a solid. Natural fibres, such as wool, are good at trapping air, providing effective insulation. Polystyrene also traps air, so it's used in walls and floors to insulate houses.

Activity – Conductors and insulators

Write the word (or words) that best completes each statement:

1. When heated, atoms vibrate *faster* .

2. Conduction is the movement of *metal particles* .

3. Metal atoms are good *conductors* .

4. A substance that holds heat is an *insulator* .

5. Insulators are good at trapping *air* .

Investigate: Conductors

Good heat conductors feel cold to touch. If you step from carpet onto lino or a tiled floor (in bare feet), the lino or tiles will feel colder than the carpet. This is because the lino or tiles quickly conduct the body heat way from your skin. On carpet, your body heat doesn't escape so easily, because carpet is an insulator that traps air. Does this mean that heat conductors have a lower temperature?

Find samples of different materials – e.g. wood, metals, cardboard, lino, polystyrene, wool, cork. Place your hand on each material and arrange them in order from coldest to warmest.

Which materials feel cooler? _____

Why is this? _____

All the materials are in fact at the same room temperature. Some materials just 'feel' cooler because they are good conductors of heat (from your hand).

Science history: Joule

James Joule (1818–1889) showed that heat was a form of energy. His name is given to the unit of energy called the joule, which equals 1 watt per second.

Thermos

Air is an excellent insulator; but an even better insulator is a place with no air – called a vacuum. Without molecules of air, there will be no heat movement. Although it's impossible to make a perfect vacuum, the thermos (vacuum flask) makes use of a pretty good near-vacuum to keep things at the same temperature – a thermos 'tries' to keep hot things hot and cold things cold.

Inside the outer casing of the thermos is a flask made of glass or plastic. The flask has a double wall with a vacuum between the layers. This reduces to a minimum heat conduction between whatever is inside the thermos and the outside. However, heat energy can pass through a vacuum – radiated as heat rays. Therefore, the thermos flask has a silver surface (like a mirror), reflecting the heat rays back to where they came from. (For a hot liquid, heat rays from the hot liquid are reflected back into

Explore, describe, and represent patterns and trends for everyday examples of physical phenomena such as movement, forces, electricity and magnetism, light, sound, waves, and heat.

the liquid, helping to keep it hot; for a cold liquid, heat rays from the outside environment are reflected back out to the environment, helping to keep the liquid inside the thermos cool.)

A thermos greatly limits heat transfer through its walls, but a little heat still escapes, especially round the cap.

Activity – Vacuum flask

Label this diagram of a thermos, using the following words:

Vacuum layer   **Cap**   **Outer casing**   **Silver surface**

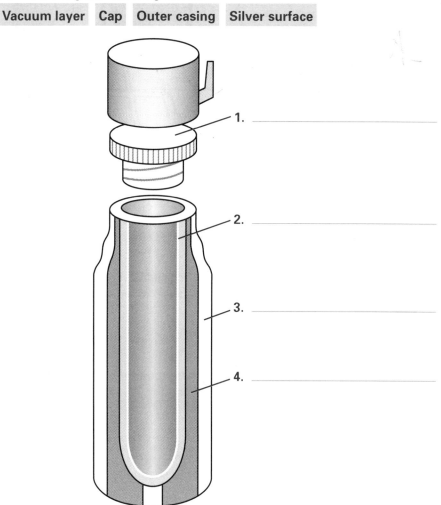

1. _____

2. _____

3. _____

4. _____

Group Activity

Find examples of insulators and conductors in the natural world. For example – animals have fur as an insulator and sweat to conduct heat from their bodies. Students illustrate an example of each.

Explore, describe, and represent patterns and trends for everyday examples of physical phenomena such as movement, forces, electricity and magnetism, light, sound, waves, and heat.

© ESA Publications (NZ) Ltd, Customer freephone: 0800-372 266

Unit 7   Electromagnetism

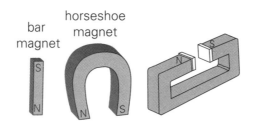

bar magnet   horseshoe magnet

Invisible attraction

Magnetism is a force found between some rocks and some metals. It's named after magnetite, a magnetic iron ore. Magnets can be found naturally, or they can be made out of metals such as iron, nickel and cobalt (magnetised by another magnet). Some materials are attracted to magnets (they are magnetic), but most others, such as plastic, are not.

The magnetic force is concentrated at the poles of a magnet. A magnet creates a force field – a pattern spreading out from the poles – as shown by the pattern of iron filings above a magnet:

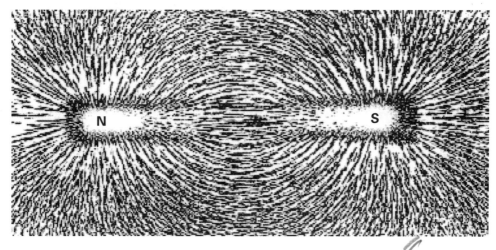

The Earth's core is a giant magnet, creating a magnetic field around our planet. One pole of a magnet will always try to and point towards the Earth's North Pole and the other towards the Earth's South Pole.

For any two magnets, opposite poles attract and same poles repel.

In a magnetic substance, such as iron or steel, each atom behaves like a tiny magnet, with its own poles. When a piece of iron is un-magnetised, the north-south poles all point in different directions. But, when fully magnetised, the atoms of iron all line up and point in the same direction.

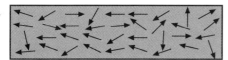

Unmagnetised

Magnetised

Magnetic substances can be magnetised by stroking them with a permanent magnet. This makes the

Explore, describe, and represent patterns and trends for everyday examples of physical phenomena such as movement, forces, electricity and magnetism, light, sound, waves, and heat.

atoms line up. A magnet can lose its magnetism when the atoms become jumbled. Demagnetising can happen when a magnet is hit or heated up.

Activity – Explaining magnetism

1. A compass has a magnetised needle suspended inside it. Explain how it works. *The magntised needle is opposite to the North pole put so it is attracted to it so the needle points that way.*

2. Explain what happens when a steel nail is stroked with a magnet. *It will become magnetised.*

3. Explain the glowing Auroras seen near the Earth's poles. _____

4. Does magnetic attraction work under water? *Yes. It passes through water.*

Electromagnetism

Magnetism is very closely related to electricity. This connection is called electromagnetism.

*   A magnet can be used to make an electric current. Wire is looped around a magnet, and the magnet is moved back and forth – making an electric current in the wire. Moving the wire while keeping the magnet still has the same effect. This is how most electricity in the world is made.

*   An electric current produces its own magnetic field. Electrons have an electric field, but when an electron moves it also creates a magnetic field. This idea is used to make electromagnets. Wire is connected to a power supply, and then looped around a rod of iron. (The iron core makes the field stronger. The coil of wire can be a temporary magnet all by itself.) The power is turned on and current moves through the wire – making the iron into a temporary magnet.

Activity – True or False?

1. A magnet produces electricity on its own.   True / **False**

2. Power stations use electromagnetism.   **True** / False

3. Electricity can be used to make magnets.   **True** / False

4. An electromagnet can not be switched off.   True / **False**

Shocking science: Landmine detectors

There are about 110 million active landmines buried throughout the world. They cause thousands of innocent deaths and injuries every year. Electromagnets are useful for finding mines. A landmine detector has a coil of wire that creates an electromagnetic field. If this field strikes any conductive material underground (e.g. the metal part of a land mine), it causes the material to generate a magnetic field. This magnetic field is then detected.

The electromagnetic spectrum

We are surrounded by a wide range of electromagnetic waves called the electromagnetic spectrum. The electromagnetic spectrum includes radio waves, microwaves, light and X-rays. These waves have different wavelengths, some long, some short. Wavelength is the distance between wave crests:

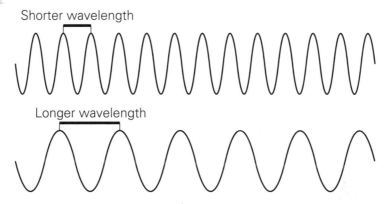

- **Gamma rays** have the smallest wavelengths – *smaller* than the width of an atom. They also have the most energy, and can kill living cells.

- **X-rays** have very short waves, about the width of an atom. Scientists usually refer to the energy of X-rays, rather than their wavelength.

- **Ultraviolet waves** are longer than X-rays but still only about the width of a virus! We can't see these. Some UV rays cause sunburn.

- **Light waves** are the visible part of the electromagnetic spectrum. We see visible light as the colour spectrum. Each colour has a different wavelength, seen separated out in a rainbow.

- **Infrared waves** have a wavelength similar to the length of microscopic organisms. Some infrared waves are thermal – it's the heat we feel from the Sun or a fire.

- **Microwaves** have wavelengths measured in centimetres. They are used to cook food, for cellphone calls, and as radar detectors.

- **Radio waves** are the longest in the electromagnetic spectrum. They range from about 30 cm up to hundreds of metres long. They can carry radio, television and cellphone signals. Antennae are needed to pick up the signals.

Group Activity

Brainstorm a list of the vocabulary, symbols and technology associated with the electromagnetic spectrum. Students choose one each to define. Assemble the definitions in a booklet.

© ESA Publications (NZ) Ltd,
Customer freephone: 0800-372 266

Explore, describe, and represent patterns and trends for everyday examples of physical phenomena such as movement, forces, electricity and magnetism, light, sound, waves, and heat.

Activity – Electromagnetic waves

Write the names of the types of wave in the boxes.

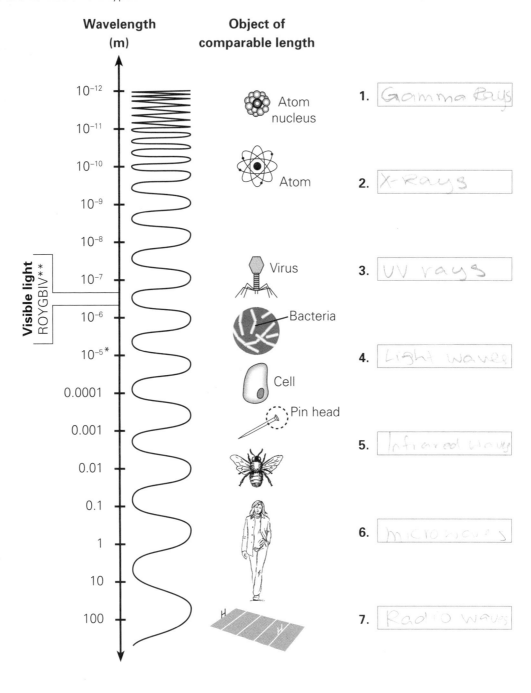

Wavelength (m)

Object of comparable length

Atom nucleus

Atom

Virus

Bacteria

Cell

Pin head

1. ~~Gamma Rays~~

2. ~~X-Rays~~

3. ~~UV rays~~

4. ~~Light waves~~

5. ~~Infrared rays~~

6. ~~microwaves~~

7. ~~Radio waves~~

$*10^{-5} = 0.00001 = \dfrac{1}{100\ 000}$

**ROYGBIV = Red, Orange, Yellow, Green, Blue, Indigo, Violet

Unit 8   Electronics R

Electric signals are controlled by Transistors, signals are used to send sound and Picture

The electronic revolution

Electronic circuits are part of most modern appliances. Electronics is a relatively recent branch of science. It started when scientists first detected electromagnetic waves. They realised these waves could be used to send signals from one place to another. The first signals transmitted were electrical pulses (Marconi used Morse code), which became known as radio.

The basis of electronics is the electric signal, which is used to carry information. Inside electronic appliances there are circuits that carry an electric current. By switching current on and off (or varying its strength), signals can be sent to operate the appliance. Electric signals are mainly controlled by devices called transistors – usually made with the element silicon (Si). Signals can be used to send sound and pictures.

A huge advance in electronics was the invention of microchips in 1971. Microchips are thin layers of silicon about 1 cm square. The ability to fit many transistors onto a single chip led to a revolution in electronics, especially in computers. A microchip sends and receives signals – electrical pulses that turn on and off at high speed.

The other big breakthrough was the use of a digital (number) code to organise the signals. 'Digital' information is anything that has been translated into a two-number code (called binary – consisting of 0 or 1). Sounds, images, and computer data can be converted into code. A digital microchip can store millions of bits of code.

Electromagnetic waves could be used to send signals Marconi used Morse code. The fo first signal sent were electric pulses known as radio

© ESA Publications (NZ) Ltd,
Customer freephone: 0800-372 266

Explore, describe, and represent patterns and trends for everyday examples of physical phenomena such as movement, forces, electricity and magnetism, light, sound, waves, and heat.

Activity – Crossword

Clues across:

3.   A new branch of science. *Electronics*

5.   Devices that control electric current. *Transistors*

7.   The man who sent one of the first radio signals. *Marconi*

8.   The microchip led to the invention of the home *Computer*

Clues down:

1.   Transistors and microchips are made of *Silicon*.

2.   Invisible waves. *Electromagnetic signal*

4.   Another word for an electrical pulse.

6.   Very small electronic device that holds many transistors. *Microchip*

Telephones

Telephones use electromagnetism. When you speak into a telephone, a microphone converts the sound waves into an electric signal (current). This is a varying signal – it constantly changes in volume and pitch. This signal is sent using land lines (wires or fibre optics) or radio waves that bounce off satellites. At the receiving phone, the signal is converted back to reproduce the sound of your voice.

Cell (mobile) phones use microchips that enable you to talk, text, take photos, and go online. They are called 'cell' phones because of the limited area (a cell) where they can operate. Each cell has a land-based aerial station. A cellphone must be within range of an aerial.

Cellphones use radio waves, land lines and microwaves:

The message from the sender's cellphone travels by radio waves to the nearest cellphone base station aerial.

↓

The cellphone base station sends the message to a telephone exchange using a land line.

↓

The telephone exchange sends the message to the cellphone base station nearest to the receiver's cellphone using microwaves.

↓

The cellphone base station nearest to the receiver's cellphone sends the message by radio waves to the receiver's cellphone.

Activity – Cellphones

Draw a diagram that shows how a cellphone call is transmitted:

Sender's Cellphone   message travels by Radio waves to nearest cellphone Base

To — Sends message ←

using land lines   ← Telephone exchange

Sends message to nearest Receivers cellphone base using Microwaves

Sends message by Radio waves to Receivers cellphone

Digital music

A huge amount of digital information can be stored on a Compact Disc (CD).

Digital music is stored on a CD using a number code similar to that used by computers. It's printed in a tight spiral track (about 5 km long!). The binary code is stamped onto the disc as a series of microscopic bumps and hollows. A laser light scans the bumpy pattern on the disc to read the code, which is then translated back into music.

Group Activity

Digital technology advances rapidly. Some say ebooks will replace printed books eventually. Survey a range of readers to find out their opinions on this issue. Debate the question:

"Do e-devices make us think better than books do?"

*Explore, describe, and represent patterns and trends for
everyday examples of physical phenomena such as movement,
forces, electricity and magnetism, light, sound, waves, and heat.*

MATERIAL WORLD
The structure of matter

Unit 1   Atoms

Atomic structure

All matter in the Universe (from planets to pencils!) is made from the same building-blocks – particles called atoms. Atoms are incredibly small – the full stop at the end of this sentence is made up of about two billion atoms.

Inside each atom are still smaller particles, called subatomic particles. There are three main kinds:

- Electrons – negatively charged.
- Protons – positively charged.
- Neutrons – no charge.

Scientists have detected even smaller particles, such as quarks. Protons and neutrons are made of quarks, but they are difficult to measure.

Inside an atom is like a mini-solar system. It's mostly empty space. Protons and neutrons clump together as a dense nucleus in the centre. The nucleus is most of the atom's mass. The nucleus is orbited by the electrons. Electrons aren't fully understood – sometimes they behave like a wave of energy around the nucleus, sometimes they behave like a particle.

Activity – Atomic particles

Label this diagram of an atom:

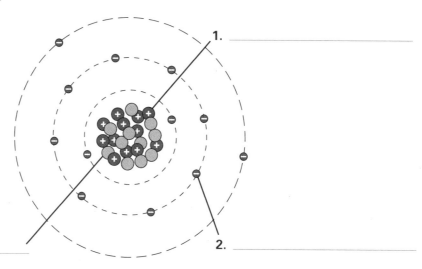

1. _____

2. _____

3. _____

Charges in atoms

Electrons and protons maintain an atom's shape through attraction between their opposite electrical charges. There are the same number of electrons and protons in an atom, so the charges are balanced.

AO: Begin to develop an understanding of the particle nature of
matter and use this to explain observed changes.

© ESA Publications (NZ) Ltd,
Customer freephone: 0800-372 266

The attraction charges in an atom are strong, but they can be broken. Atoms can be split by striking the nucleus with another particle – this is what happens in a nuclear reaction inside a nuclear power reactor.

Activity – Subatomic particles

Fill in this table that lists the three main parts of an atom.

Subatomic particle	Where is it found?	Kind of charge

World's largest experiment

There is still a lot that scientists don't understand about atomic particles. The biggest physics experiment in the world is being conducted in order to try to find out more about the smallest known particles. It's a scientific instrument called the **Large Hadron Collider** (**LHC**) in a 27-km-long circular tunnel in Europe. The LHC is a huge particle accelerator that fires beams of subatomic particles (such as protons) in opposite directions around the tunnel. When the particles collide with each other, scientists hope to discover new particles. The collisions should also give information about how the Universe began just after the Big Bang.

Ernest Rutherford: New Zealand scientist
who first determined the structure of the atom.

Group Activity

Read a range of scientific texts (digital or printed) that have diagrams, pictures or models of atoms and atomic structures. Compare the images and evaluate them. Students comment on which ones they find most helpful and clear. Students construct their own models of atoms and particles and identify the best features of one another's models.

© ESA Publications (NZ) Ltd,
Customer freephone: 0800-372 266

AO: Begin to develop an understanding of the particle nature of matter and use this to explain observed changes.

Unit 2   Elements

There are different kinds of atom, containing different numbers of subatomic particles. A substance containing atoms with the same number of protons is called an **element**. For example, all the atoms in the element carbon are the same – each carbon atom contains 6 protons. (Each carbon atom also has 6 electrons, but the number of neutrons varies.)

Atomic number is the number of protons in the nucleus of each atom. Atoms of each element always have the *same* atomic number.

Elements are also given a **mass number** – the number of protons plus neutrons. Because the number of neutrons can vary, mass numbers are *not* usually whole numbers.

Elements are grouped and arranged by their atomic number in a grid called the Periodic Table.
Elements are listed on the Periodic Table as **symbols**. There are 112 elements currently listed on the Periodic Table – more are being discovered (or made). The most common elements in the Universe are hydrogen (H) and helium (He).

The element iron appears in the Periodic Table as:

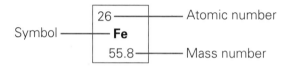

Symbol ——— 26 ——— Atomic number
Fe
55.8 ——— Mass number

Each atom of iron has 26 protons.

The symbol of iron is Fe (from the Latin word for iron, *ferrum*).

The mass number is 55.8, made up of:

- 5.8% iron atoms with 28 neutrons

- 91.7% iron atoms with 30 neutrons

- 2.2% iron atoms with 31 neutrons

- 0.3% iron atoms with 32 neutrons.

$_6$ H^1, He^2

Carbon
Nitrogen ^{7}
oxygen ^{8}
oxygen

Copper Cu
Sodium Na
Iron Fe

*AO: Begin to develop an understanding of the particle nature of
matter and use this to explain observed changes.*

© FSA Publications (NZ) Ltd,
Customer freephone: 0800-372 266

Activity – Atomic maths (1)

Atomic Number

1	
H	
1.0	Atomic Mass

Periodic Table:

1	2											13	14	15	16	17	18
																	2 **He** 4.0
3 **Li** 6.9	4 **Be** 9.0											5 **B** 10.8	6 **C** 12.0	7 **N** 14.0	8 **O** 16.0	9 **F** 19.0	10 **Ne** 20.2
11 **Na** 23.0	12 **Mg** 24.3	3	4	5	6	7	8	9	10	11	12	13 **Al** 27.0	14 **Si** 28.1	15 **P** 31.0	16 **S** 32.1	17 **Cl** 35.5	18 **Ar** 40.0
19 **K** 39.1	20 **Ca** 40.1	21 **Sc** 45.0	22 **Ti** 47.9	23 **V** 50.9	24 **Cr** 52.0	25 **Mn** 54.9	26 **Fe** 55.9	27 **Co** 58.9	28 **Ni** 58.7	29 **Cu** 63.6	30 **Zn** 65.4	31 **Ga** 69.7	32 **Ge** 72.6	33 **As** 74.9	34 **Se** 79.0	35 **Br** 79.9	36 **Kr** 83.8
37 **Rb** 85.5	38 **Sr** 87.6	39 **Y** 88.9	40 **Zr** 91.2	41 **Nb** 92.9	42 **Mo** 95.9	43 **Tc** 98.9	44 **Ru** 101	45 **Rh** 103	46 **Pd** 106	47 **Ag** 108	48 **Cd** 112	49 **In** 115	50 **Sn** 119	51 **Sb** 122	52 **Te** 128	53 **I** 127	54 **Xe** 131
55 **Cs** 133	56 **Ba** 137	71 **Lu** 175	72 **Hf** 179	73 **Ta** 181	74 **W** 184	75 **Re** 186	76 **Os** 190	77 **Ir** 192	78 **Pt** 195	79 **Au** 197	80 **Hg** 201	81 **Tl** 204	82 **Pb** 207	83 **Bi** 209	84 **Po** 210	85 **At** 210	86 **Rn** 222
87 **Fr** 223	88 **Ra** 226	103 **Lr** 262	104 **Rf** 261	105 **Db** 262	106 **Sg** 263	107 **Bh** 264	108 **Hs** 265	109 **Mt** 268									

Lanthanide Series

57 **La** 139	58 **Ce** 140	59 **Pr** 141	60 **Nd** 144	61 **Pm** 147	62 **Sm** 150	63 **Eu** 152	64 **Gd** 157	65 **Tb** 159	66 **Dy** 163	67 **Ho** 165	68 **Er** 167	69 **Tm** 169	70 **Yb** 173

Actinide Series

89 **Ac** 227	90 **Th** 232	91 **Pa** 231	92 **U** 238	93 **Np** 237	94 **Pu** 239	95 **Am** 241	96 **Cm** 244	97 **Bk** 249	98 **Cf** 251	99 **Es** 252	100 **Fm** 257	101 **Md** 258	102 **No** 259

Use the Periodic Table to answer the following questions.

Hint: There are the same number of electrons and protons in an atom.

1. What is the atomic number of copper (Cu)? _____

2. What is the mass number of krypton (Kr)? _____

3. How many protons does an oxygen atom have in its nucleus? _____

4. How many electrons does sodium (Na) have in an atom? _____

Activity – Atomic maths (2)

1. What is the total number of protons and neutrons in the element gold $^{79}_{197}$**Au**? _____

2. How many neutrons in an atom of phosphorus $^{15}_{31}$**P**? _____

3. For a titanium $^{22}_{48}$**Ti** atom, list the number of protons, neutrons and electrons:

_____ , _____ , _____

4. In a uranium $^{92}_{238}$**U** atom, list the number of protons, neutrons and electrons:

_____ , _____ , _____

© ESA Publications (NZ) Ltd,
Customer freephone: 0800-372 266

AO: Begin to develop an understanding of the particle nature of matter and use this to explain observed changes.

The Periodic Table

The Periodic Table groups elements together because they have similar properties. The elements are arranged both in rows (across the table) and in columns (down the table). The main groups are metals and non-metals.

- The metals share similar properties – e.g. they are shiny, and are good electrical conductors.
- Non-metals are dull and poor conductors.

In between these two groups are the semi-metals, which have properties of both. For example, silicon is shiny, yet it is an inefficient conductor and is brittle.

Activity – Elements and the Periodic Table

On the Periodic Table on the previous page:

- Colour the non-metals – He, C, N, O, F, Ne, P, S, Cl, Ar, Br, Kr, Xe, Rn – one colour (e.g. blue).
- Colour the semi-metals – B, Si, Ge, As, Se, Sb, Te – another colour (e.g. green).
- Colour the metals – the rest of the Table – another colour (e.g. yellow).

Science history: Pattern discovered

In 1869, the Russian chemist Dmitri Mendeleyev worked out that elements form a pattern of properties when they are grouped according to their atomic number. Mendeleyev's original Periodic Table revolutionised chemistry.

Activity – Name the element

Use the Periodic Table on the previous page to help to identify the following:

1. Atomic number is 2 and is lighter than air: _Helium_   He
2. A metal with the symbol Zn: _Zinc_   Zn
3. Has 20 protons and is found in bones and teeth: _Calcium_   Ca
4. The most common gas on Earth; it has 7 protons and usually 7 neutrons: _Nitrogen_   N
5. The second most abundant element in the Earth's crust; it has 14 protons: _Silicon_   Si
6. A lightweight metal with mass number 27.0: _Aluminium_   Al

AO: Begin to develop an understanding of the particle nature of
matter and use this to explain observed changes.

© ESA Publications (NZ) Ltd,
Customer freephone: 0800-372 266

Properties and changes of matter

Unit 3   Compounds and mixtures

Compounds

When two or more different elements are joined together they form a compound. For example, sodium (Na) and chlorine (Cl) combine to form the compound sodium chloride (NaCl), which is salt. A compound is made by a **chemical** reaction. It's usually difficult to separate a compound into its original elements. Compounds have properties that differ from those of the original elements. For example, in solid salt, Cl was originally a green poisonous gas and Na was originally a light metal.

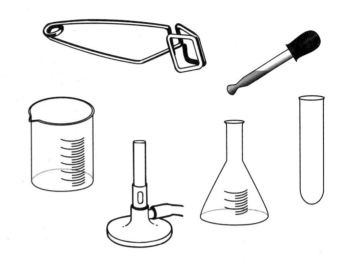

Each compound has a formula, e.g. NaCl for salt. The formula tells the number of atoms (or ratio) of each element in the compound. For example, the compound H_2O (pure water) has a basic structure of 2 hydrogen atoms joined to 1 oxygen atom. Compounds are pure substances. Only a few common substances are compounds.

Mixtures

Most common substances in the world are mixtures (e.g. drinks, air, petrol). Even tap water is a mixture, because the water that makes up the tap water has many other substances dissolved in it. A mixture is made up of at least two compounds. The parts of a mixture are usually much more easily separated than are the atoms that make up a compound. For example, distillation readily separates water from salt in seawater.

Mixtures are impure substances.

Activity – Element, compound or mixture?

Write these substances in the correct group:
Nitrogen (N), Carbon dioxide (CO_2), Soil, Glucose ($C_6H_{12}O_6$), Tin (Sn), Milk, Silver (Ag), Methane (CH_4), Sea water.

Element	Compound	Mixture

AO: Group materials in different ways, based on the observations and measurements of the characteristic chemical and physical properties of a range of different materials.

Group Activity

The teacher sets up stations around the room with different compounds and mixtures. Students move around the stations and try to identify them visually. Then devise tests for each to prove if it's a compound or mixture.

Separating mixtures

Mixtures can be separated into the original substances they are made of. The separation method depends on the physical properties of the substances. Physical properties include density, particle size, boiling point, solubility, magnetism, and colour.

Density

Gold is more dense than other minerals, so it settles to the bottom during gold panning. The water can then be decanted (poured off).

A centrifuge can speed up the settling of solids. It spins liquids to speed up the rate at which the more dense particles sink.

Size

Filtering is a separation method that traps solid particles and allows liquid to flow through. The type and size of particles trapped depends on the size of pores in the filter (or holes in the sieve).

Boiling point

Distilling is a process that separates a mixture of liquids. One of the liquids will have a lower boiling point. When the mixture is heated, this liquid evaporates first and its vapour can be collected. The process can be repeated for the liquid with the next lowest boiling point.

Solubility

Evaporation is a method used to separate the solids dissolved in a solution, such as salt from water.

AO: Group materials in different ways, based on the observations and measurements of the characteristic chemical and physical properties of a range of different materials.

© ESA Publications (NZ) Ltd, Customer freephone: 0800-372 266

Activity – Methods of separation

Draw a line to link each substance with the correct method of separation. You may need to do some research.

Substance to separate	Method used
Salt from seawater	Magnetic attraction
Coffee grounds from liquid	Filtering
Iron from aluminium	Centrifuging
Alcohol from water	Decanting
Blood cells from blood plasma	Distillation
Coloured dyes from black ink	Evaporation
Sand from water	Chromatography

Investigate: Distillation

Distillation can separate out a solid that is dissolved in water.

Method

1. Make a salt solution by stirring 2 tablespoons of salt into a cup of water.

2. Pour the solution into a small pot and heat. Put a large lid, at an angle, on the pot.

3. Heat the solution until boiling.

4. Place a bowl under the lower edge of the pot lid to catch the drips.

Results

The water condenses on the lid and drips off. Taste it when cool. Where is the salt?

Your blood filter

Your kidneys are a built-in filter for your blood. The kidneys are small but energetic – processing about 180 litres of blood a day! Waste is filtered from blood as it's forced through the kidneys. The waste liquid (filtrate) is urine.

Unfortunately, the kidneys can be seriously damaged by diabetes. One type of diabetes is associated with being overweight. So, look after your kidneys by taking care with your diet.

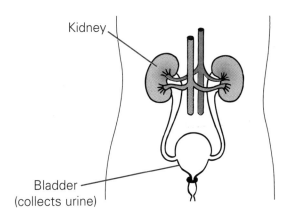

Kidney

Bladder
(collects urine)

© ESA Publications (NZ) Ltd,
Customer freephone: 0800-372 266

AO: Group materials in different ways, based on the observations and measurements of the characteristic chemical and physical properties of a range of different materials.

Unit 4   Acids and bases ℞

Acids

Acids are sour (the word 'acid' comes from the Latin *acidus*, meaning 'sharp'). Weak acidic solutions (e.g. vinegar) are relatively safe. Strong acids are highly corrosive – enough to burn your skin off (e.g. sulfuric acid, H_2SO_4).

Bases

Bases are bitter and soapy, such as bicarbonate of soda (baking soda). Weak bases are used in baking. Some relatively weak bases can be dangerous – such as ammonia solution (NH_4OH). Ammonia is used in household cleaning because it reacts with grease and is toxic to micro-organisms. Strong bases are hazardous and corrosive – such as caustic soda solution, NaOH.

Neutral substances

Neutral substances are neither acid or base – for example, distilled water.

What makes an acid or base?

All acids contain hydrogen. When added to water, an acid splits apart; the hydrogen forms hydrogen ions, H^+. It is these hydrogen ions which make a solution acidic.

Bases are compounds that react with hydrogen ions. Bases that dissolve in water (called **alkalis**) produce hydroxide ions, OH^-. Hydroxide ions react with hydrogen ions, H^+.

Neutral substances have an equal amount of hydrogen ions and hydroxide ions.

When an acid is added to a base, the hydrogen ions, H^+, from the acid, react with the hydroxide ions, OH^-, from the base, to form water, H_2O:

$H^+ + OH^- \rightarrow H_2O$

When a base is added to an acid, the acid is said to be neutralised.

When an acid is added to a base, the base is said to be neutralised.

AO: Group materials in different ways, based on the observations and measurements of the characteristic chemical and physical properties of a range of different materials.

*© ESA Publications (NZ) Ltd,
Customer freephone: 0800-372 266*

Activity – Explaining acids and bases

1. Why does lemon juice sting? *Weak acid But sharp taste*

    What acid is in lemon juice? *citric acid*

2. Why is ammonia a good cleaner of dirty surfaces? *ammonia reacts with acids greasy foods are acidic and washed away with water, It is also Toxic kills germs*

3. Why is hydrogen important in acids? *Hydrogen forms Hydrogen ions that make a solution acidic*

4. What makes a base an alkali? *When bases dissolve in water become alba*

5. Why is tap water not neutral? *Because water is not Pure, Tap water in cities is slitly slitly Alkaline, stored rainwater in tanks is slitly acidic because it desolves carbon dioxide (CO₂) in it*

Indicators

To test if a substance is an acid or a base, you can use an indicator. Indicators contain pigments that are sensitive to acidity.

Universal indicator is made up of several pigments:

-   Acidic solutions turn universal indicator red (strong acid), orange (mild acid) or yellow (weak acid).
-   Bases turn universal indicator blue (mild alkali) or purple (strong alkali).
-   Neutral substances turn universal indicator green.

Indicators can be made from plants that contain pigments sensitive to acids and bases – such as the juice of red cabbage, blackberries and elderberries.

Experiment: Testing substances

Method

Finely chop half a red cabbage and add to half a litre of boiling water. Soak for half an hour. Strain off the purple solution – this is your indicator. (Set some aside for later experiments – *Making acid* in this Unit and *Soap pH* in Unit 5).

You need 10 jars or test tubes. Pour a little indicator into each jar. Keep some indicator aside as a control. Add a spoonful of each substance listed in the table that follows.

Compare the colour to the indicator control.

© ESA Publications (NZ) Ltd,
Customer freephone: 0800-372 266

AO: Group materials in different ways, based on the observations and measurements of the characteristic chemical and physical properties of a range of different materials.

Results

Substance	Acid	Base	Neutral
Citric acid			
Baking soda			
Tea			
Vinegar			
Tap water			
Distilled water			
Laundry powder			
Fruit juice			
Ammonia (bleach)			
Egg white			
Tartaric acid			
Alcohol			

Conclusions

What seemed to be the strongest acid? _____

What seemed to be the strongest base? _____

Which substances were neutral? _____

AO: Group materials in different ways, based on the
observations and measurements of the characteristic chemical
and physical properties of a range of different materials.

Activity – pH scale

The pH scale shows the strength of an acid or base, on a scale from 0 to 14. Acids have a pH below 7, bases have a pH above 7, neutral solutions have a pH of exactly 7.

Following are some common pH values:

Ammonia pH 11.9, Baking soda pH 8.4, Blood pH 7.4, Eggs pH 7.9, Fruit juice pH 3.5, Hydrochloric acid pH 0.5, Lemon juice pH 2.3, Milk pH 6.5, Pure alcohol pH 7, Saliva pH 6.7, Sodium hydroxide pH 14, Stomach acid pH 1, Tomato juice pH 4.1, Vinegar pH 3, Yoghurt pH 5.

Write the substances on the pH scale. Label the scale with 'Strong acid, Weak acid, Neutral, Weak base, Strong base.'

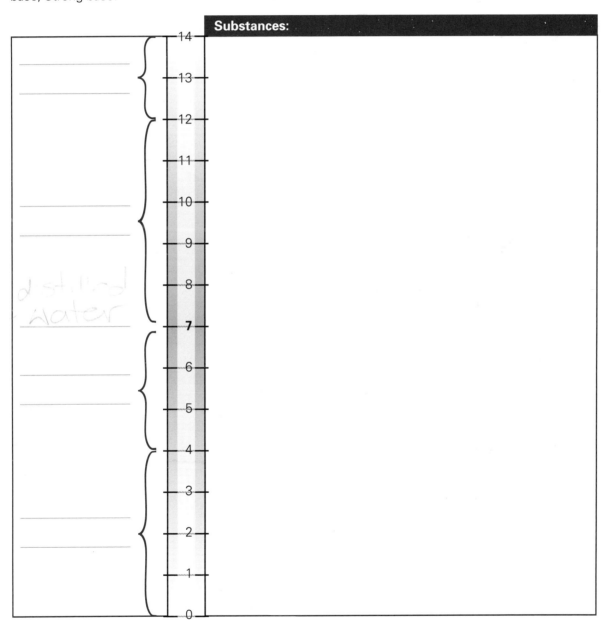

Substances:

AO: Group materials in different ways, based on the observations and measurements of the characteristic chemical and physical properties of a range of different materials.

Investigate: Making acid

This experiment uses the red cabbage indicator from the previous experiment. Place a drinking straw in the indicator and blow bubbles into the solution. Observe any colour change.

What happens? _____

_____

Why? Your breath contains CO_2. The CO_2 gas reacts with the water:

$$CO_2 + H_2O \rightarrow H_2CO_3$$

H_2CO_3 is carbonic acid. Carbonic acid lowers the pH of the indicator solution.

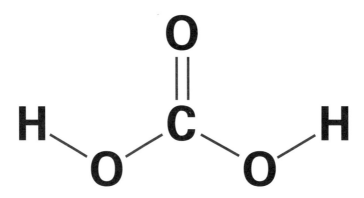

Structure of carbonic acid

Group Activity

Class learns the chemical formulae for common acids and bases. Create a memory game to reinforce learning.

AO: Group materials in different ways, based on the observations and measurements of the characteristic chemical and physical properties of a range of different materials.

© ESA Publications (NZ) Ltd, Customer freephone: 0800-372 266

Unit 5   Soap chemistry R

How is soap made?

The pH of soap is between 8.5 and 10, therefore soap is a base. Soap is made from an alkali (base) substance – usually sodium hydroxide, NaOH. The other main ingredient is animal fat or vegetable oil. Early settlers in New Zealand often used wood ashes (an alkali) and whale fat to make soap. The fats or oils react with the alkali to form soap.

How does soap work?

Soap molecules consist of two distinct parts:

- A 'tail' – a long chain of carbon atoms with hydrogen atoms attached to them. The tail is hydrophobic (it 'hates' water); it is, however, attracted to grease and oil.

- A 'head' – made up of a carbon atom and two oxygen atoms that overall have a negative charge. The head is hydrophilic (it 'loves' water).

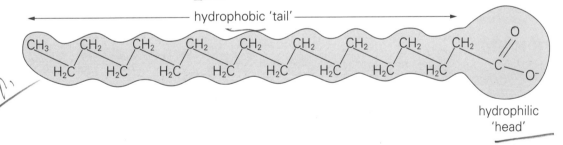

Imagine washing greasy hands with soap. The tail and head end of the soap molecule are pulling in opposite directions. The hydrophobic tail buries itself in the grease, while the hydrophilic head pulls towards the water – the combination of which means the grease particles are lifted off your hands. The hydrophobic ends continue to attract the grease particles and keep them suspended in water. This helps to wash the grease away.

The hydrophobic ends also push water molecules apart at the surface of a body of water. This breaks the surface tension of the water. Detergents (which act like soap) are added to cleaning sprays so the spray spreads out on its target when it hits its target (rather than sitting on the target as raised individual droplets).

AOs: Begin to develop an understanding of the particle nature of matter and use this to explain observed changes. Relate the observed, characteristic chemical and physical properties of a range of different materials to technological uses and natural processes.

Activity – Soap Quiz

Choose the correct word to fill in the gaps:

Soap has a pH that is **1.** _____ (above/below) seven; therefore it is classified as a(n)

2. _____ (acid/ base) substance. Soap can be made from **3.** _____ (fat/water) mixed

with NaOH. NaOH is an **4.** _____ (acid/alkali).

Carbon chains with **5.** _____ (hydrogen/helium) attached are found in soap. Hydrophilic

means water **6.** _____ (hating/loving), while hydrophobic means water **7.** _____

(hating/loving). The hydrophobic end of a soap molecule attaches easily to **8.** _____

(grease/liquid). Soap disturbs water **9.** _____ (suspension/molecules) at the surface of a

body of water.

Investigate: Soap pH

Use the red cabbage indicator from Unit 4.

Poor some indicator solution onto a bar of white soap.

Colour change? _____

Pour some indicator onto your hands and allow it to dry. Wash your hands in soapy water.

Colour change? _____

Explain the colour changes: _____

Hard water

No Foam   *Soft H₂O*   *Foam with Soap*

Water can be hard or soft. Hard water has calcium and/or magnesium ions in it. Hard water creates many difficulties for humans. It's not easy to wash with – a lot more soap is needed to get a good lather (foam). The calcium/magnesium ions react with the soap and form a scum. Some soap is used up by bonding with the ions – this reduces the amount of soap available for cleaning.

AOs: *Begin to develop an understanding of the particle nature of matter and use this to explain observed changes. Relate the observed, characteristic chemical and physical properties of a range of different materials to technological uses and natural processes.*

Another problem with hard water is that it causes the scum inside electric jugs, chalky deposits inside water pipes, and the dirty ring around the bath-tub.

All hard water can be softened using washing powder. Washing powder contains a water-softening chemical called sodium carbonate (Na_2CO_3). The carbonate part bonds with calcium/magnesium ions, removing them so that the soap part of the washing powder remains to clean the clothes.

Activity – Hard water

Name three household problems resulting from hard water.

1. _____

2. _____

3. _____

Investigate: Testing water hardness

Select two different water sources – e.g. a sample of water from your house tap, a sample of bottled water. Test the water samples with liquid soap first (not laundry soap).

Method

Measure equal amounts of each water into two jars.

Add a drop of liquid soap to each jar and shake. Keep adding a drop at a time and shake. Count the drops, and keep adding until the water forms a lather of bubbles.

Which water sample was harder? _____

What was the source of the 'harder' water? _____

Why might the harder water source have more calcium/magnesium ions? _____

Group Activity

Students bring samples of a wide variety of soaps. As a class, devise a 'fair test' of the soap samples.

- How will cleaning ability be assessed?
- How will the dirt samples be kept fair?
- Can liquid soaps and solid soaps be fairly compared?
- What other factors apart from the cleaning ability of soap are important to consumers?

AOs: Begin to develop an understanding of the particle nature of matter and use this to explain observed changes. Relate the observed, characteristic chemical and physical properties of a range of different materials to technological uses and natural processes.

Chemistry and society

Unit 6   Pollution (1) R

Acid rain

Causes

Acid rain is a type of pollution. It is created when acidic gases pollute the air. These gases come from factories, power stations, and vehicle exhausts, as well as natural sources such as volcanoes.

When fossil fuels (e.g. oil, petrol, diesel) are burned, they produce carbon dioxide. The CO_2 gas dissolves in water vapour in the air, and forms a weak acid called carbonic acid. $CO_2 + H_2O \rightarrow H_2CO$

Coal-fired power stations cause a lot of acid rain. Burning coal that contains sulfur (and most coal contains sulfur) produces sulfur dioxide – another acidic gas – which reacts with water vapour to produce sulfurous acid. (Sulfurous acid is about ten times stronger than carbonic acid.)

Effects

Acid rain damages plants, animals, and humans. It soaks into soils and takes away nutrients so they are not available to trees, as well as causing direct damage to the surface of leaves. Acidic soil has caused widespread food crop failures. Acid rain in waterways makes fish and other water creatures unhealthy.

Acid rain can irritate human skin, and cause breathing problems such as asthma and bronchitis. Acid rain corrodes stone buildings (especially those made of limestone or cement), and metal structures.

Activity – Acid city

Acid rain is an environmental problem in populated areas of Europe, Asia, and North America.

List the causes and effects of acid rain – use the table following this picture.

Causes	Effects

Activity – Acid lakes

Normal lake and river pH is about 6. In North America, many lakes have a pH below 5 because of acid rain pollution. At pH 5.5, water snails begin to die out. At pH 5.0, fish eggs will not hatch. At pH 4.5, trout will die. At pH 4.0, frogs are the only animals still alive. Below pH 4.0, no animals survive in lakes.

Use this information to create a pH scale showing damage to living things.

pH	Effects
6–8	
5.5	

AOs: Compare physical and chemical changes. Relate the observed, characteristic chemical and physical properties of a range of different materials to technological uses and natural processes.

pH	Effects
5.0	
4.5	
4.0	
Below 4.0	

Investigate acid rain

1.   Rainwater

Measure the acidity of rainwater where you live. Collect rainwater and test it with Universal Indicator Paper (available from a pharmacy) or Universal Indicator solution.

2.   Corrosion

Place a piece of chalk (limestone) in a jar of vinegar (weak acid), and another piece in a jar of water. After 3 days, compare the amount of corrosion. Repeat with a copper coin.

What do you observe? _____

Fighting acid rain

Reducing use of car travel cuts down on CO_2 emissions. New cars are fitted with a catalytic converter which cleans exhaust gases. Making electricity from wind and hydropower reduces the amount of harmful gases released in the atmosphere. Recycling plastics prevents acid rain because fossil fuels are burned in the production of plastic packaging.

AOs: *Compare physical and chemical changes. Relate the observed, characteristic chemical and physical properties of a range of different materials to technological uses and natural processes.*

© ESA Publications (NZ) Ltd, Customer freephone: 0800-372 266

Activity – Take action

List three things you could do to prevent acid rain.

1. _____

   _____

   _____

2. _____

   _____

   _____

3. _____

   _____

Group Activity

Debate the pollution issue.

- Is pollution inevitable in a functioning city?

- Can all pollution be prevented?

- What value do we place on caring for the natural world?

Look at both sides of the issue, weighing up lifestyle, economics, environment and values.

© ESA Publications (NZ) Ltd,
Customer freephone: 0800-372 266

AOs: Compare physical and chemical changes. Relate the observed, characteristic chemical and physical properties of a range of different materials to technological uses and natural processes.

Unit 7   Pollution (2)

Oil

Oil is a fossil fuel, made when plant and animal remains are compressed underground for millions of years. Crude (raw) oil is a mixture of compounds called hydrocarbons, made up of hydrogen and carbon atoms. Humans rely on oil as the main fuel in the world for industry and transport, and to make electricity, plastics and fertilisers. Unfortunately, oil pollutes the environment.

Oil pollution comes from vehicles, factories, oil tanker spills, and pipeline leaks. About half of the oil in the oceans comes from waste oil lost from vehicles (carried in water run-off from the land). Oil is difficult to get rid of – it doesn't dissolve in water (oil and water molecules repel each other), and it floats on the surface (it's less dense than water).

Oil spills poison fish and shellfish, and harm marine mammals and bird life. This has a flow-on effect to fisheries, coastal industries, and tourism.

Activity – Thinking

1.   What alternatives to oil are there for transport?

2.   How might seabirds be harmed by oil?

3.   What does oil do in sea water?

AO: Relate the observed, characteristic chemical and physical properties of a range of different materials to technological uses and natural processes.

Ocean oil clean-up

1. Soaking

Sorbents are materials that soak up liquids. Oil sorbents include woodchips, polystyrene and straw. A floating barrier of straw bales can help to protect a harbour entrance from an oil spill.

2. Skimming

Skimmers remove oil from the surface. A skimming machine pulled by a boat can suck oil off the surface.

3. Dispersing

Chemical dispersants can be sprayed onto oil spills to break the oil up. They work a bit like dish-washing detergent. Molecules of dispersant attach to oil molecules, breaking the oil into droplets that can then mix with sea water. However, chemical dispersants can also be toxic to life.

Activity – Make a dispersant

Pour a cup of water into a screw-top jar. Add a half cup of cooking oil. Put the lid on and shake well. Allow the mixture to settle. Observe. Now add a spoonful of dishwashing detergent and shake.

What happens to the oil? _____

Water pollution

Water is polluted by sewage, petrol, farm waste, and leachate (seepage) from rubbish dumps. Many New Zealand waterways (lakes and rivers) are seriously contaminated by intensive agriculture – including fertiliser run-off and animal wastes. Contamination by nitrogen fertilisers can cause excess nutrients in rivers, resulting in choking growth of plants.

Human populations discharge sewage into waterways and into the ocean. For example, some towns pump untreated sewage into the sea, often making beaches dangerous for swimming because of toxic bacteria.

Chemicals such as pesticides and herbicides are washed by rain into waterways. In cities, chemicals such as cleaning products and oils are washed down drains into waterways and harbours. You can protect water by not tipping chemicals down the drain. Oil and paints should be disposed of at a recycling centre.

AO: Relate the observed, characteristic chemical and physical properties of a range of different materials to technological uses and natural processes.

Activity – Polluted crossword

Clues:

1. As this industry becomes more intensive, it's likely to cause more pollution.

2. A word that involves 'pollution' by chemicals.

3. Pollution that seeps out of rubbish dumps.

4. Places with high human population, so pollution is worse.

5. Never tip oil and chemicals down these.

6. Sewage should always be _____ .

7. This chemical/nutrient causes excessive plant growth in rivers.

Shocking science: War pollution

The 1991 Gulf War was a disaster for the environment. Millions of barrels of oil were released into the Persian Gulf, forming a slick 50 km × 12 km in size. Approximately 67 million tonnes of oil were burned as well. The black clouds turned day into night and produced toxic acid rains.

AO: Relate the observed, characteristic chemical and physical properties of a range of different materials to technological uses and natural processes.

© ESA Publications (NZ) Ltd,
Customer freephone: 0800-372 266

Unit 8   Food safety

Chemicals in our food

Are there chemicals in our food? "Yes" – food is made up entirely of chemicals – hundreds of thousands of them! Most of these are useful chemicals, such as proteins, carbohydrates, fats, and many nutrients. But there are also small amounts of other chemicals in foods, some of which can harm us. Some foods contain chemicals that are **natural toxins** that can harm humans. Sometimes, chemicals, such as preservatives and flavourings, are *added* to foods by people. Sometimes, there are also small amounts of **agricultural chemicals**, such as insecticides, left on our food. To stay healthy, it's important to know about what is in our food.

Natural toxins

Some food plants contain natural toxins which the plant may use to protect itself – for example, against insects. These can make humans sick if we eat them. Following are some common natural toxins to be aware of.

- Apricot kernels (pips) – contain a chemical that can be dangerous if consumed.
- Kumara – can produce a toxin that makes it taste bitter. Remove damaged parts of a kumara before cooking.

103

AO: Relate the observed, characteristic chemical and physical properties of a range of different materials to technological uses and natural processes.

- Potatoes – contain toxins, especially in the sprouts and green skin.
- Kidney beans – raw beans contain toxins – but these can be broken down by first soaking, then boiling at high temperature until cooked.
- Rhubarb – oxalic acid is mostly in the leaves, so don't eat them.

Activity – Toxins

Create a poster that illustrates the naturally toxic parts of some vegetables and fruits and gives advice on how to safely prepare these vegetables and fruits for eating.

Food additives

Additives include preservatives, colours, acidity regulators, antioxidants, emulsifiers, stabilisers, thickeners, anticaking agents, humectants and flavour enhancers. These are chemicals used to make food last longer and to modify its flavour or appearance. Preservatives are widely used additives that can stop food going off or from drying out. Some additives are natural chemicals, such as salt – which is both a preservative and a flavouring agent. Many processed foods have artificial additives in them. For example, aspartame is a manufactured chemical used instead of natural sugar. Nutrients are also sometimes added to food, especially if they are lacking in our diet, such as iodine (a mineral) which is added to salt. Additives in food are tested to make sure they are safe at the level used (although some people are allergic to additives).

Salt and pepper are common flavour enhancers

Activity – Additive research

Find out the function of each of these additives. Use the government's New Zealand Food Safety Authority (nzfsa) website.

1. acidity regulator: _adjust Acid or Alkaline level_
2. antioxidant: _Prevent Fats and oil Breaking, Slows Flavour and colour chan_
3. emulsifier: _____
4. stabiliser: _Keeps ingredients mixed_
5. anticaking agent: _Promotes free flow of Particles (e.g. salt)_
6. humectant: _Prevents food from drying out_

AO: Relate the observed, characteristic chemical and physical properties of a range of different materials to technological uses and natural processes.

© *ESA Publications (NZ) Ltd,*
Customer freephone: 0800-372 266

Activity – Food labels

The law requires that all food labels list the additives used in a product (including additive class and code). Make a survey of commonly used food products. Create a chart to record the information. Group additives using their class name and code numbers. Look for patterns and trends on the chart. Which additives are used most? Which are natural? Which products contain the most additives?

Agricultural chemicals

Pesticides can be extremely toxic to handle and protective clothing may have to be worn

Agricultural chemicals are widely used to produce plants and animals. They include fertilisers, insecticides, herbicides, fungicides, and veterinary medicines. These chemicals often leave small amounts (residues) in the final food product that we eat. New Zealand law sets limits on the residues in our food to minimise the health risks. This system relies on food producers obeying the law and using the legal amounts of chemicals. Every five years there is survey of food to sample the levels of agricultural chemical contamination. The surveys have shown that over half our food has residues, and that grain and meat are more likely to have pesticide residues. People who want to avoid artificial chemical residues often eat only organic food.

Group Activity

Debate the use of agricultural chemicals compared with the organic farming system. Consider cost, pest and weed control, yields, environmental effects, human diet.

© ESA Publications (NZ) Ltd,
Customer freephone: 0800-372 266

AO: Relate the observed, characteristic chemical and physical properties of a range of different materials to technological uses and natural processes.

ANSWERS

Answers are not given where student responses will vary.

LIVING WORLD

Unit 1   Plant Kingdom

Name game   (page 1)

Plant word	Feature of plant
trifolium	Deeply serrated leaves
odorata	Very small leaves
serrata	Leaves grouped in threes
microphylla	Plant spreads close to ground
prostrata	Named after Cunningham
cunninghamii	Plant has a smell

Plant types   (page 2)

1. Produce spores   **2.** Conifers   **3.** Flowering plants   **4.** Monocots

Plant classification   (page 2)

1. Monocot.   **2.** Dicot.

Adapting   (page 4)

Answers will vary. Examples:

1. Pine needles are thin leaves – to reduce moisture/heat loss.

2. Alpine flowers are bright to attract scarce insects.

3. Cacti have fast-growing, surface roots.

4. Succulents store water in thick, fleshy stems.

5. Orchids attract specific moths as pollinators.

Research   (page 4)

Nepenthes rajah has a large, jug-shaped leaf that holds 3.5 litres of water – it can trap and digest frogs, lizards, rats, and birds.

Unit 2   New Zealand forests

New Zealand time line   (page 5)

1. Seaweed-like plants.   **2.** Mosses, ferns, conifers.   **3.** Podocarps.   **4.** Flowering plants.

© ESA Publications (NZ) Ltd,
Customer freephone: 0800-372 266

Forest species (page 7)

Answers could include:

Conifer – kauri, kaikawaka (mountain cedar).

Podocarp – kahikatea, totara, rimu, matai, miro.

Broadleaf – rata, tawa.

Beech – silver beech, red beech, mountain beech, black beech.

Climber – supplejack, clematis, jasmine, rata vine, bush lawyer.

Natural processes (page 8)

Answers could include:

1. Tree roots absorb rain; the forest floor acts like a sponge; water is released slowly from forests.

2. Leaves take in CO_2, the CO_2 combines with water to make food for the tree.

3. Birds need trees to nest in; they need food such as leaves, berries, nectar and insects.

Unit 3   Bird evolution

Seagull (page 10)

Suggested answers:

1. Curved wing top.   **2.** Flexible wing.   **3.** Light beak.

4. Feathers.   **5.** Large breastbone.   **6.** Lightweight skeleton.

Wings and arms (page 10)

1. Humerus.   **2.** Radius.   **3.** Phalanges.   **4.** Ulna.

Unit 4   Protists

True or False? (page 14)

1. False   **2.** True   **3.** False   **4.** False   **5.** True   **6.** False

The malaria cycle (page 14)

1.   **C** Plasmodium injected into human's bloodstream.

2.   **E** Plasmodium multiplies in person's red blood cells.

3.   **B** Plasmodium released into human's bloodstream.

4.   **D** Plasmodium in human blood sucked up by another mosquito.

5.   **A** Plasmodium multiplies in mosquito.

Unit 5   Viruses

Deeper thinking (page 18)

1. By a worldwide programme of vaccination.

2. A billionth of a metre.

3. To produce copies of the virus once inside a host cell.

4. By contact; by sneezing tiny droplets containing the virus.

© ESA Publications (NZ) Ltd,
Customer freephone: 0800-372 266

5. High-density living means quicker spread, air travel takes viruses around the world in as little time as a couple of days.

Stages of attack (page 19)

1. Virus lands on bacteria cell.
2. Virus injects its genes.
3. Bacteria makes copies of viral genes.
4. Viral copies break out of host cell.

Flu research (page 20)

Influenza type	Years	Death toll (estimated)
Spanish flu	1918	50 million
Asian flu	1957–1958	1 million
Hong Kong flu	1968	0.75 million

Unit 6 Seeds

Reproductive cycle (page 22)

1. Pollination.
2. Fertilisation.
3. Fruiting.
4. Dispersal.
5. Germination.

Seed containers (page 22)

Answers will vary:

1. Apple.
2. Apricot.
3. Strawberry.
4. Bean.
5. Pumpkin.
6. Corn.
7. Walnut.
8. Wheat.

Seed adaptations (page 24)

Species	Seeds dispersed by	Seed adaptation
Sycamore	Wind	Wing
Scotch thistle	Wind	Fluff/down
Hook grass	Animals	Hooks/barbs
Coconut	Sea water	Waterproof
Pond iris	Water	Floating
Broom	Explosion	Dry pods

Unit 7 Digestive system

The journey of food (page 25)

1. Mouth.
2. Salivary glands.
3. Oesophagus.
4. Stomach.
5. Duodenum.
6. Ileum.
7. Large intestine.
8. Rectum.

Healthy food pyramid: True or False? (page 27)

1. False
2. True
3. True
4. False
5. True

PLANET EARTH AND BEYOND

Unit 1   The Universe

Universe quiz: True or False? (page 28)

1. False.   **2.** True.   **3.** False.   **4.** True.   **5.** True.   **6.** True.

The Universe time line (page 29)

Time	Event
0	A tiny fireball suddenly explodes and expands.
billionth billionth billionth of the first second	Fireball expands trillions of times; intense energy and particles created.
3 minutes	Hydrogen, helium, lithium formed; cooling.
300 000 years	Light escapes; Universe clears to transparent space; gravity gathers elements.
300 million years	Stars and galaxies forming, Universe expanding.
9 billion years	Milky Way formed.
10 billion years	First life on Earth.
14 billion years	Present day; Universe still expanding.

Deeper thinking (page 30)

1. Einstein said that the gravitational field of an object will distort space and bend light. Black holes proved him right.
2. Because space and time interact closely. If one is warped, so is the other.
3. X-rays can't get through Earth's atmosphere – satellites carrying telescopes are needed. Satellites travel above Earth's atmosphere.

Unit 2   The planets

Planet table (page 35)

	Temperature (°C)	Climate conditions	Atmospheric gases	What is the surface made of	Day length
Mercury	−180 to +450	Very hot and very cold	Thin	Rock	58 Earth days
Venus	+464	Hot and cloudy	CO_2, sulfuric acid	Rock	243 Earth days
Mars	−143 to +30		CO_2	Rock and soil	1 Earth day
Jupiter	−110	Large hurricanes	Hydrogen, helium	Gases	9.93 hours
Saturn	−140	Very strong winds	Ammonia	Gases	10.66 hours
Uranus	−197		Methane		

	Temperature (°C)	Climate conditions	Atmospheric gases	What is the surface made of	Day length
Neptune	–200	Extreme winds	Hydrogen, helium, methane	✕	✕

Unit 3   Tectonics

Earth layers   (page 36)

Layer	Inner core	Outer core	Lower mantle	Upper mantle	Crust
Composition	*Metal*	Metal	Rock	Rock	Rock
Solid or Liquid	Solid	*Liquid*	Solid, but plastic	Solid	Solid
Thickness (km)	1 200	2 250	*2 900*	100	8–70
Temperature (°C)	5 000	4 000	3 500	*1 000*	Cool

Tectonics   (page 37)

Suggested answers:

1. rock.

2. heated rock.

3. plates.

4. plate movement.

5. boundaries.

6. San Andreas.

7. continental.

8. subduction.

New Zealand plates   (page 38)

1. Australian-Indian Plate.

2. Southern Alps.

3. Subduction zone.

4. Pacific Plate.

Unit 4   Earthquakes

The Richter scale   (page 39)

Note that the effect of a quake will vary with factors such as depth and type of land.

Magnitude on Richter scale	Effects and damage caused by earthquake
3.0–3.9	Usually can be felt, but rarely causes damage.
4.0–4.9	Felt inside, light shaking of objects and some noise. No significant damage.
5.0–5.9	A moderate shake which can cause significant damage to poorly built structures but not to well-built ones.
6.0–6.9	A strong shake which can be very damaging up to about 150 km from the epicentre.
7.0–7.9	A major quake which can cause severe damage across a wide area.
8.0–8.9	Such quakes occur about once a year somewhere in the world and are extremely destructive over a wide area.

Quake damage   (page 40)

1. Buildings collapsed, fires, some land pushed up and some land subsided, 256 people killed.

2. Broken gas pipes caused fires which burned the old city, 500+ people killed.

© ESA Publications (NZ) Ltd,
Customer freephone: 0800-372 266

3. Tsunami across several oceans, at least 283 000 people killed.

4. Over 300 000 killed, followed by a cholera outbreak.

5. Over 16 000 killed and massive tsunami damage.

Unit 5   Mountain building

Mountain records   (page 42)

Mountain	Continent	Height (m)
Everest	Asia	8 848
K2	Asia	8 611
Aconcagua	South America	6 960
Kilimanjaro	Africa	5 965
Matterhorn	Europe	4 478

Deeper thinking: Wearing down   (page 44)

1. Heating expands and cooling shrinks the outer layer of rock, which eventually peels off.

2. When water freezes, it expands.

3. Acidic water dissolving limestone.

4. Tree roots make slopes stable by binding soil, stones and other material together – removing trees causes landslides.

5. In glaciated areas of the Southern Alps, such as Tasman, Fox, Franz Josef.

Unit 6   Climate change

Thinking about ice ages   (page 45)

1. Water is frozen into ice sheets and glaciers – means less sea water.

2. Plants and animals would have to adapt to the cold, colonise areas now becoming suitable, or become extinct.

Global warming   (page 46)

1. Reflected.   2. Radiated.   3. Absorbed.   4. Re-radiated.

CO_2 levels   (page 47)

1.

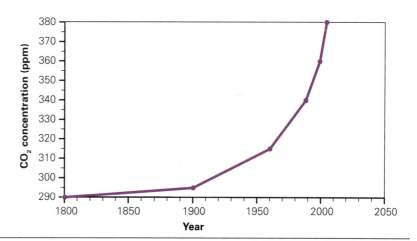

© ESA Publications (NZ) Ltd,
Customer freephone: 0800-372 266

2. The levels climb more rapidly in the last 30 years.

3. As population has increased hugely in the last 40 years, so have industry and power generation, farming, etc.

Sea level (page 48)

1. Christchurch, Auckland, Dunedin.

2. New York, London, Tokyo.

3. Flooding damage, permanent displacement of people, damage to businesses.

Unit 7   Climate

Climate terminology (page 50)

1. Temperate.   **2.** Latitude.   **3.** Continental.   **4.** Climate.

5. Sea level.   **6.** Equator.   **7.** Tropics.

Climate zones examples (page 51)

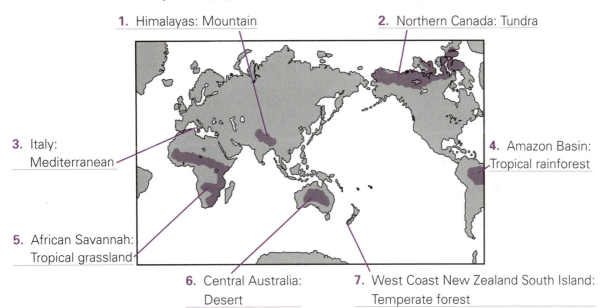

1. Himalayas: Mountain

2. Northern Canada: Tundra

3. Italy: Mediterranean

4. Amazon Basin: Tropical rainforest

5. African Savannah: Tropical grassland

6. Central Australia: Desert

7. West Coast New Zealand South Island: Temperate forest

Unit 8   Meteorology

Weather map symbols (page 53)

1. High.   **2.** Low.   **3.** Isobars.   **4.** Cold front.   **5.** Warm front.

Read a weather map (page 54)

1. E   **2.** C   **3.** F   **4.** B   **5.** A   **6.** D

© ESA Publications (NZ) Ltd,
Customer freephone: 0800-372 266

PHYSICAL WORLD

Unit 1   Motion

Movements (page 55)

1. movement   **2.** orbit   **3.** air   **4.** gravity   **5.** directions

Galileo's legendary experiment (page 56)

Large and small masses fall at the same rate! Gravity pulls all objects with an equal force. Two weights dropped will reach the ground at the same time. It shouldn't matter if they are different sizes or weights, as long as they are the same shape.

Speed units (page 57)

1. 3 mm/s   **2.** 1.5 m/s   **3.** 25 km/h   **4.** 10 m/s   **5.** 300 km/h
6. 1 000 km/h   **7.** 1 224 km/h   **8.** 30 000 km/h   **9.** 300 000 km/s

Explaining momentum (page 58)

1. The person has momentum too – they will continue moving forward unless they hold on to some part of the bus.
2. The heavy ball gathers momentum as it swings, and when it hits the brick wall, the momentum of the ball is passed to the bricks, which fly apart.
3. The meteorite had a huge amount of momentum, which pushed aside a large amount of earth.

Unit 2   Gravity

Gravity (page 60)

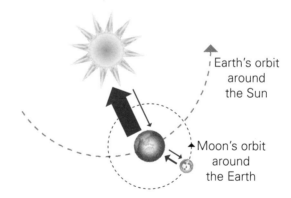

Earth's orbit around the Sun

Moon's orbit around the Earth

Explaining mass and weight (page 60)

1. Six times higher than you can on Earth. This is because the gravity on the Moon is only 1/6 that of gravity on Earth.
2. a. The same as that on Earth.   **b.** One-sixth that on Earth.
3. Air resistance slows it.
4. Ten times your mass.

Unit 3   Floating and sinking

Floating and sinking definitions (page 61)

1. Upthrust: Force that pushes upwards against objects in a liquid.

© *ESA Publications (NZ) Ltd,*
Customer freephone: 0800-372 266

2. Displacement: Water pushed aside by an object.
3. Floating: When displaced, water creates enough upthrust to support an object.
4. Sinking: Not enough upthrust to support an object.
5. Density: Mass of a substance (of a given volume).

Research (page 63)

1. Fish maintain their buoyancy with an air-filled sac called a swim-bladder. If a fish wants to sink, it compresses the air in the bladder.
2. A submarine alters its buoyancy by emptying or filling special tanks. Compressed air is used to empty the tanks of water, decreasing the submarine's density so it rises or floats; water from the outside flooding into the tanks increases the submarine's density, so the submarine sinks.
3. Answers could include: wetsuit, weight belt, air-tank, backpack.

   Wetsuits are much less dense than water, so when a diver puts on a wetsuit, their overall buoyancy increases. To enable them to dive under, divers must use weights to reduce their buoyancy. Weights include weighted belts, the weight of air-tanks, and backpack weights.

Unit 4   Energy

Kinetic or potential? (page 64)

Example	Kinetic energy	Potential energy
Bird flying fast near the ground	✓	
Heat from a fire	✓	
Skydiver in a plane		✓
Lightning flash	✓	
Petrol in a tank		✓
Bird flying high and slowly		✓
Rolling soccer ball	✓	
Pizza		✓
Stretched rubber band		✓

Energy transfer (page 65)

Stored chemical energy (Food energy) from muscles → Kinetic energy (Movement energy) of bat → Kinetic energy (Movement energy) of ball → Kinetic energy (Movement energy) and Gravitational Potential energy (Height energy) in ball's flight → Kinetic energy (Movement energy) and Sound energy and Heat energy from the broken window

Energy chain (page 66)

1. heat *and* light.
2. chemical *energy.*
3. kinetic *energy and* heat.

Unit 5   Renewable energy

Solar limitations   (page 68)

Sunlight is variable and changes with the weather (sunny/cloudy), the seasons (e.g. summer vs winter), location and time of day. Sunlight is not concentrated in one place, so a large collector is needed to get a lot of energy.

Other sources of renewable energy   (page 69)

Suggested answers:

Energy source	Advantage	Disadvantage
Nuclear power	Does not produce greenhouse gases such as those from burning coal	Dangerous waste, difficult to dispose of
Solar power	Pays for itself eventually, and is renewable	Relatively small power output; needs sunny days
Wind power	Renewable	Takes up a lot of space; needs windy days
Hydroelectric power	Little pollution	Floods a lot of land

Unit 6   Heat

Conductors and insulators   (page 70)

Answers may vary.

1. faster   **2.** heat energy through a solid   **3.** conductors
4. insulator   **5.** air

Vacuum flask   (page 72)

1. Cap.   **2.** Silver surface.   **3.** Outer casing.   **4.** Vacuum layer.

Unit 7   Electromagnetism

Explaining magnetism   (page 74)

1. The North-seeking pole of the suspended needle (the South magnetic pole of the compass needle) is attracted to the North Pole of the Earth. The needle turns to point north.
2. The magnetic domains align themselves in chains which leaves 'free' N and S poles at each end of the nail – the nail becomes a magnet.
3. Charged particles from the Sun are directed by the Earth's magnetic field towards the Poles, where they ionise the upper atmosphere – as the ions de-excite, they emit the light seen.
4. Yes, magnetic force passes through water and many other non-magnetic materials.

True or False?   (page 74)

1. False   **2.** True   **3.** True   **4.** False

Electromagnetic waves (page 76)

1. Gamma rays.
2. X-rays.
3. Ultraviolet waves.
4. Light waves.
5. Infrared waves.
6. Microwaves.
7. Radio waves.

Unit 8 Electronics

Crossword (page 78)

1. Silicon
2. Electromagnetic
3. Electronics
4. Signal
5. Transistors
6. Microchip
7. Marconi
8. Computer

Cellphones (page 79)

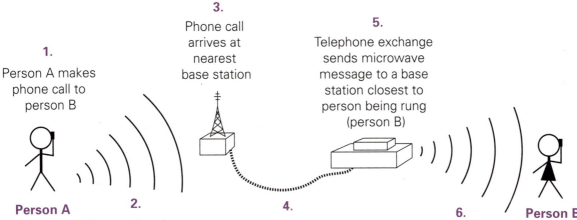

1.
Person A makes phone call to person B

Person A

2.
Person A's voice travels as a radio wave

3.
Phone call arrives at nearest base station

4.
Base station transmits message through an underground phone cable to a telephone exchange building

5.
Telephone exchange sends microwave message to a base station closest to person being rung (person B)

6.
Base station sends radio waves to person B's cellphone

Person B

MATERIAL WORLD

Unit 1 Atoms

Atomic particles (page 80)

1. Protons (in nucleus). **2.** Electrons. **3.** Neutrons (in nucleus).

Subatomic particles (page 81)

Subatomic particle	Where is it found?	Kind of charge
Electron	Orbiting nucleus	Negative
Proton	In nucleus	Positive
Neutron	In nucleus	No charge

Unit 2 Elements

Atomic maths (1) (page 83)

1. 29 **2.** 83.8 **3.** 8 **4.** 11

Atomic maths (2) (page 83)

1. 197 **2.** 16
3. 22 protons, 26 neutrons, 22 electrons **4.** 92 protons, 146 neutrons, 92 electrons

Name the element (page 84)

1. Helium. **2.** Zinc. **3.** Calcium.
4. Nitrogen. **5.** Silicon. **6.** Aluminium.

Unit 3 Compounds and mixtures

Element, Compound or mixture? (page 85)

Element	Compound	Mixture
Nitrogen (N), Tin (Sn), Silver (Ag).	Carbon dioxide (CO_2), Glucose ($C_6H_{12}O_6$), Methane (CH_4).	Soil, Milk, Sea water.

Methods of separation (page 87)

Substance to separate	Method used
Salt from seawater	Magnetic attraction
Coffee grounds from liquid	Filtering
Iron from aluminium	Centrifuging
Alcohol from water	Decanting
Blood cells from blood plasma	Distillation
Coloured dyes from black ink	Evaporation
Sand from water	Chromatography

Unit 4 Acids and bases

Explaining acids and bases (page 89)

1. It's a weak acid with a sharp taste. Lemon juice contains citric acid.
2. Ammonia reacts with acids (greasy foods are acidic) so the acids can be washed away with water. Ammonia is also toxic, killing micro-organisms.
3. Hydrogen forms hydrogen ions, H^+; it is H^+ ions that make a solution acidic.
4. The base is soluble – the solution has lots of hydroxide ions, OH^-.
5. Tap water is not pure – it has other chemicals dissolved in it. Tap water in towns and cities is usually slightly alkaline, because of the chemicals added to it. Tap water in the country from stored rainwater (e.g. a rainwater tank) is usually slightly acidic, because rain has CO_2 dissolved in it.

pH scale (page 91)

Strong acids (spans pH 0–3.9) Fruit juice, Hydrochloric acid, Lemon juice, Stomach acid, Vinegar
Weak acids (spans pH 4–6.9) Milk, Saliva, Tomato juice, Yoghurt
Neutral (pH 7) Pure alcohol
Weak bases (spans pH 7.1–11.9) – Ammonia, Baking soda, Blood, Eggs
Strong bases (spans pH 12–14) – Sodium hydroxide

Unit 5 Soap chemistry

Soap Quiz (page 94)

1. above	**2.** base	**3.** fat	**4.** alkali	**5.** hydrogen
6. loving	**7.** hating	**8.** grease	**9.** molecules	

Hard water (page 95)

1. Soap lathers less effectively, use more soap.
2. Scum forms around bath when soap is used for washing.
3. Chalky deposits clog hot water pipes.

Unit 6 Pollution (1)

Acid city (page 96)

Suggested answer:
Causes: Smoke or gas emissions from factories, power stations and vehicles.
Effects: Damage to crops, animals, human health; buildings corrode.

Acid lakes (page 97)

pH	Effects
6–8	Normal lake life
5.5	Water snails die
5.0	Fish eggs die
4.5	Trout die

pH	Effects
4.0	Only frogs able to survive
Below 4.0	All animals die

Take Action (page 99)

Answers could include: walking or cycling instead of taking a car, using public transport, not using plastic bags offered to customers for carrying items from shops, supporting alternative energy sources (wind, solar power).

Unit 7 Pollution (2)

Thinking (page 100)

Answers will vary.
1. Biofuels, electric cars and trains, cycling, walking.
2. Oil clogs feathers which prevents birds flying, swimming, and looking for food; also poisons them and they can go into shock.
3. Oil floats on top of sea water and doesn't mix with sea water.

Polluted crossword (page 102)

1. Agriculture
2. Contamination
3. Leachate
4. Cities
5. Drains
6. Treated
7. Nitrogen

Unit 8 Food safety

Additive research (page 104)

1. acidity regulator: adjusts acid or alkaline levels; gives a sour or sharp taste; slows growth of microbes.
2. antioxidant: helps to prevent fats and oils breaking down; slows flavour loss and colour change.
3. emulsifier: keeps oil- based and water-based ingredients mixed together.
4. stabiliser: keeps ingredients mixed that don't usually stay mixed.
5. anticaking agent: promotes free-flow of particles (e.g. salt).
6. humectant: prevents food from drying out.

INDEX

acceleration 55–7

acid rain 96–8, 102

acids 27, 88–92

adaptations 3–4, 9–10, 14, 23–5

agricultural chemicals 103, 105

agriculture 101

air pressure 52–3

air resistance 9, 59

algae 13–14

Amoeba 14

Alpine Fault 41

alpine plants 3

Andes 37

Andromeda galaxy 28

anticyclones 52

Aoraki *see* Mount Cook

Archaeopteryx 9

Archimedes' Principle 62

aspartame 104

asteroids 31

atmosphere (of Earth) 8, 31, 46–7, 52, 98

atmosphere (of planets) 33–5

atomic number 82–4

atomic structure 80–1

auroras 54, 74

Australian-Indian Plate 38

bacteria 14, 18–19, 25, 76, 101

bacteriophage 18–19

bases 88–92

beaks 9–12

Beaumont, William 27

beech trees 1, 6–7

Big Bang theory 29–30, 81

binary code 79

bird evolution 9–12

bird flu virus 20

black holes 29–30, 33

bones 9, 84

buoyancy 63

cacti 2–3

capsid 17–18

carbohydrates 26, 103

carbon footprint 48

carbonic acid 92, 96

carnivorous plants 4

catalytic converters 106

CD (compact disc) 79

cellphones 75, 78–9

Celsius thermometer scale 54

centrifuge 86

Chernobyl nuclear explosion 69

ciliates 14

climate change 45–7

climate zones 50–1

coal 46, 67, 69, 96

comets 31

compounds 86–8, 100

computers 77–9

conductors (electrical) 84

conductors (heat) 70–2

conifers 1, 5–7

continental climate 50

corrosion 98

crust (of Earth) 36–7, 39, 42, 68, 84

Cygnus X-1 30

dams 68

Darwin, Charles 12

density 61–2, 86

depressions (weather) 52

diabetes 87

diatoms 14

dicotyledons ('dicots') 10

diet 26–7, 87, 104

digestive system (human) 25–7

digital information 77, 79

dinosaurs 9

dispersants (chemical) 101

distillation 87

dynamics 55–6

earthquakes 37–41

Einstein, Albert 29–30, 66

electric current 74, 77–8

electromagnetism 73–6, 78

electronic circuits 77–8

electrons 67, 74, 80, 82–3

elements 37, 82–4

energy changes 65–6

enzymes 25

epicentre of earthquake 39

epiphytes 6

equator 49–50

© ESA Publications (NZ) Ltd,
Customer freephone: 0800-372 266

© ESA Publications (NZ) Ltd,
Customer freephone: 0800-372 266

© ESA Publications (NZ) Ltd,
Customer freephone: 0800-372 266